KT-234-560

00002065

IN 25 VOLUMES

*Each title on a specific drug or drug-related problem*

| | |
|---|---|
| ALCOHOL | ***Alcohol*** *and Alcoholism* |
| | ***Alcohol*** *Customs & Rituals* |
| | ***Alcohol*** *Teenage Drinking* |
| HALLUCINOGENS | ***Flowering Plants*** *Magic in Bloom* |
| | ***LSD*** *Visions or Nightmares* |
| | ***Marijuana*** *Its Effects on Mind & Body* |
| | ***Mushrooms*** *Psychedelic Fungi* |
| | ***PCP*** *The Dangerous Angel* |
| NARCOTICS | ***Heroin*** *The Street Narcotic* |
| | ***Methadone*** *Treatment for Addiction* |
| | ***Prescription Narcotics*** *The Addictive Painkillers* |
| NON-PRESCRIPTION DRUGS | ***Over-the-Counter Drugs*** *Harmless or Hazardous?* |
| SEDATIVE HYPNOTICS | ***Barbiturates*** *Sleeping Potion or Intoxicant?* |
| | ***Inhalants*** *Glue, Gas & Sniff* |
| | ***Quaaludes*** *The Quest for Oblivion* |
| | ***Valium*** *The Tranquil Trap* |
| STIMULANTS | ***Amphetamines*** *Danger in the Fast Lane* |
| | ***Caffeine*** *The Most Popular Stimulant* |
| | ***Cocaine*** *A New Epidemic* |
| | ***Nicotine*** *An Old-Fashioned Addiction* |
| UNDERSTANDING DRUGS | ***The Addictive Personality*** |
| | ***Escape from Anxiety and Stress*** |
| | ***Getting Help*** *Treatments for Drug Abuse* |
| | ***The Mood Modifiers*** |
| | ***Teenage Depression and Drugs*** |

# NICOTINE

THE ENCYCLOPEDIA OF PSYCHOACTIVE DRUGS

# NICOTINE

## An Old-Fashioned Addiction


JACK E. HENNINGFIELD, Ph.D.

*The Johns Hopkins University School of Medicine*

Adapted by Martin Jarvis M.A., B.Sc., M.Phil.



GENERAL EDITOR (U.S.A.)

Professor Solomon H. Snyder, M.D.

*Distinguished Service Professor of Neuroscience, Pharmacology, and Psychiatry at The Johns Hopkins University School of Medicine*

---

GENERAL EDITOR (U.K.)

Professor Malcolm H. Lader, D.Sc.,Ph.D., M.D., F.R.C. Psych.

*Professor of Clinical Psychopharmacology at the Institute of Psychiatry, University of London, and Honorary Consultant to the Bethlem Royal and Maudsley Hospitals*

**Burke Publishing Company Limited**

*LONDON*

First published in the United States of America 1985

This edition first published 1985


**Acknowledgements**

The author gratefully acknowledges the efforts of Judy Murphy and Robert Hutchings at the U.S. Government Office on Smoking and Health for providing detailed information and literature on the causes and effects of cigarette smoking.

Photos courtesy of AP/Wide World Photos, the American Cancer Society, the American Lung Association, the American Heart Association, *Vogue, Time, The New Yorker, The New York Times,* and the American Tobacco Company.

**CIP data**
Henningfield, Jack E.
Nicotine—(Encyclopedia of psychoactive drugs)
1. Nicotine 2. Nicotine—Physiological effect
I. Title II. Series
615'.78 RM666.T6
ISBN 0 222 01215 3 Hardbound
ISBN 0 222 01216 1 Paperback

Burke Publishing Company Limited
Pegasus House, 116–120 Golden Lane, London EC1Y OTL, England.
Printed in the Netherlands by Deltaprint Holland.

# CONTENTS

*In 1977 Britain's Health Education Council placed this advertisement in the major national newspapers to warn smokers that recently introduced cigarettes containing a mixture of cellulose and tobacco were essentially no less dangerous than regular cigarettes.*

# INTRODUCTION

The late twentieth century has seen the rapid growth of both the legitimate medical use and the illicit, non-medical abuse of an increasing number of drugs which affect the mind. Both use and abuse are very high in general in the United States of America and great concern is voiced there. Other Western countries are not far behind and cannot afford to ignore the matter or to shrug off the consequent problems. Nevertheless, differences between countries may be marked and significant: they reflect such factors as social habits, economic status, attitude towards the young and towards drugs, and the ways in which health care is provided and laws are enacted and enforced.

Drug abuse particularly concerns the young but other age groups are not immune. Alcoholism in middle-aged men and increasingly in middle-aged women is one example, tranquillisers in women another. Even the old may become alcoholic or dependent on their barbiturates. And the most widespread form of addiction, and the one with the most dire consequences to health, is cigarette-smoking.

Why do so many drug problems start in the teenage and even pre-teenage years? These years are critical in the human life-cycle as they involve maturation from child to adult. During these relatively few years, adolescents face the difficult task of equipping themselves physically and intellectually for adulthood and of establishing goals that make adult life worthwhile while coping with the search for personal identity, assuming their sexual roles and learning to come to terms with authority. During this intense period of growth

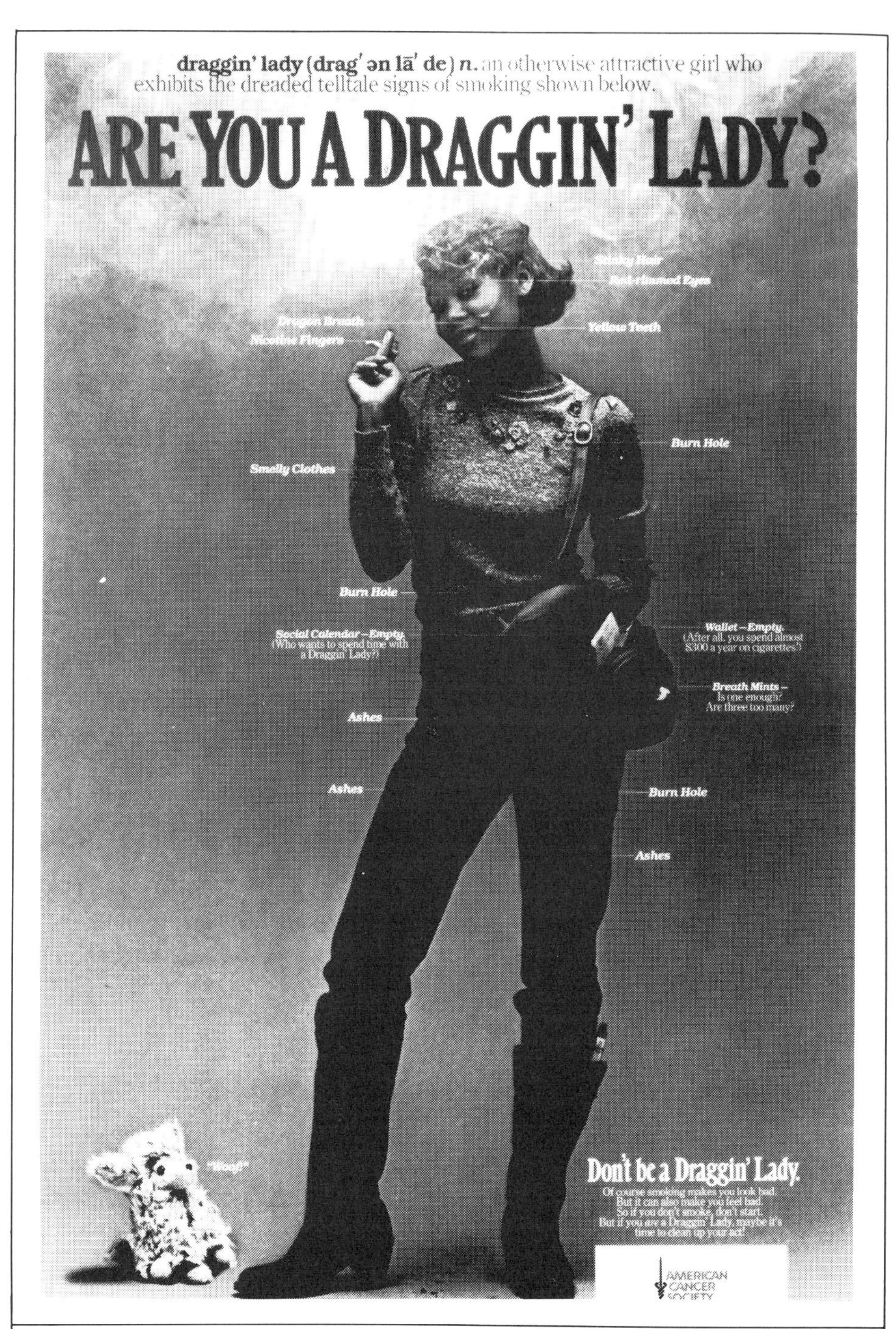


***An American Cancer Society poster cleverly emphasizes some of the negative aspects of smoking.***

and activity, bewilderment and conflict are inevitable, and peer pressure to experiment and to escape from life's apparent problems becomes overwhelming. Drugs are increasingly available and offer a tempting respite.

Unfortunately, the consequences may be serious. But the penalties for drug-taking must be put in perspective. Thus, addicts die from heroin addiction but people also die from alcoholism and even more from smoking-related diseases. Also, one must separate the direct effects of drug-taking from those indirectly related to the life-style of so many addicts. The problems of most addicts include many factors other than drug-taking itself. The chaotic existence or social deterioration of some may be the cause rather than the effect of drug abuse.

Drug use and abuse must be set into its social context. It reflects a complex interaction between the drug substance (naturally-occurring or synthetic), the person (psychologically normal or abnormal), and society (vigorous or sick). Fads affect drug-taking, as with most other human activities, with drugs being heavily abused one year and unfashionable the next. Such swings also typify society's response to drug abuse. Opiates were readily available in European pharmacies in the last century but are stringently controlled now. Marijuana is accepted and alcohol forbidden in many Islamic countries; the reverse obtains in most Western countries.

The use of psychoactive drugs dates back to prehistory. Opium was used in Ancient Egypt to alleviate pain and its main constituent, morphine, remains a favoured drug for pain relief. Alcohol was incorporated into religious ceremonies in the cradles of civilization in the Near and Middle East and has been a focus of social activity even since. Coca leaf has been chewed by the Andean Indians to lessen fatigue; and its modern derivative, cocaine, was used as a local anaesthetic. More recently, a succession of psychoactive drugs have been synthesized, developed and introduced into medicine to allay psychological distress and to treat psychiatric illness. But, even so, these innovations may present unexpected problems, such as the difficulties in stopping the long-term use of tranquillisers or slimming-pills, even when taken under medical supervision.

*The Encyclopedia of Psychoactive Drugs* provides information about the nature of the effects on mind and body of

*An American Cancer Society poster asks a question designed to prevent young nonsmokers from ever starting and to encourage young smokers to stop before they become heavily addicted.*

alcohol and drugs and the possible results of abuse. Topics include where the drugs come from, how they are made, how they affect the body and how the body deals with these chemicals; the effects on the mind, thinking, emotions, the will and the intellect are detailed; the processes of use and abuse are discussed, as are the consequences for everyday activities such as school work, employment, driving, and dealing with other people. Pointers to identifying drug users and to ways of helping them are provided. In particular, this series aims to dispel myths about drug-taking and to present the facts as objectively as possible without all the emotional distortion and obscurity which surrounds the subject. We seek neither to exaggerate nor to play down the complex topics concerning various forms of drug abuse. We hope that young people will find answers to their questions and that others—parents and teachers, for example—will also find the series helpful.

The series was originally written for American readers by American experts. Often the problem with a drug is particularly pressing in the USA or even largely confined to that country. We have invited a series of British experts to adapt the series for use in non-American English-speaking countries and believe that this widening of scope has successfully increased the relevance of these books to take account of the international drug scene.

This volume discusses the very important topic of nicotine and cigarette smoking. The account was originally written by Jack Henningfield but has been extensively modified by Martin Jarvis. The evidence that tobacco use is a form of drug addiction is carefully presented and treatments to help addicts give up are suggested. The magnitude of the public health problems presented by cigarette smoking makes this behaviour the most important of all the drug addictions.

**M.H. Lader**

*Every year cigarettes kill more Americans than were killed in World War I, the Korean War, and Vietnam combined; nearly as many as died in battle in World War II. Each year cigarettes kill five times more Americans than do traffic accidents. Lung cancer alone kills as many as die on the road. The cigarette industry is peddling a deadly weapon. It is dealing in people's lives for financial gain.*

—from an address by the late United States Senator
Robert F. Kennedy of New York
at the First World Conference on Smoking and Health
New York, September 11, 1967

*An American Cancer Society poster. The public health lobby often hires advertising agencies to communicate the anti-smoking message, taking advantage of a medium also exploited fully by the tobacco industry.*

# AUTHOR'S PREFACE

Throughout the industrialized world, there are many millions of people who smoke cigarettes. In Britain nearly forty per cent of all adults are smokers, and the proportion is similar in the United States and other developed countries. These are remarkable statistics considering that most people believe that smoking is harmful to their health, and most smokers say they would give it up if they knew of an effective way.

Why do so many people smoke cigarettes? Is smoking a "voluntary pleasure" as claimed by the tobacco industry? A behavioural habit like chewing gum? A form of self-treatment for nervousness and weight control? Or a form of drug abuse? There are elements of truth in all of these explanations. Some, however, are more valid than others.

There is now a wealth of evidence about the harmful effects of smoking. It is clear that if people smoked less, tobacco-related diseases and deaths would decrease. So it is very important to understand why people smoke.

The study of the medical effects of smoking began in earnest in the 1950s, and the first United Kingdom report on Smoking and Health by the Royal College of Physicians was published in 1962, soon to be followed by the first Surgeon General's Report in the United States. But the study of why people smoke, of the *causes* of smoking, did not begin until much later. Most of it has taken place in the last ten years.

There are several reasons why the study of the causes of smoking lagged so far behind the study of its effects. For one thing, researchers could not justify spending time and money on investigating the causes until it was shown that the effects were harmful. Also, the reaons for smoking are not easy to pinpoint. One of the major questions is whether there are physical and psychological factors that cause or influence people to smoke—factors they are not even aware of. Researchers cannot just ask smokers why they smoke. Their answers may be true as far as they go, but they are probably only a part of the story. To study a complicated behaviour pattern like smoking you need sophisticated research techniques, and such techniques have not been available until recently.

Research into causes was also delayed because of an early and rather simple-minded view of smoking. This model held that smoking was a voluntary act and that if people knew it was harmful, they would give it up. Government and health organisations made a serious effort to convince people that smoking was harmful. They publicized statistics on deaths and diseases caused by tobacco, and films of the ravages of lung cancer were shown on television. Some people gave up or cut down, but the majority carried on

***A root-like malignant cancer cell seen through an electron microscope. Smoking is a contributory factor in 30% of all deaths from cancer in the United States.***

smoking. It is still the case now that less than half of those who have ever smoked cigarettes give up permanently before the age of sixty. This provides a clue to the nature of smoking. Persisting with behaviour that is known to be damaging is a hallmark of drug abuse.

This book presents the most up-to-date thinking and scientific knowledge about the physical and psychological effects of smoking, the reasons why people smoke, and treatments to help them give up smoking. Only through a better understanding of smoking behaviour will it be possible to develop more effective ways of helping those who want to stop smoking.

***Aversion therapy exaggerates the more unpleasant aspects of an activity normally considered pleasurable. Here smoke is blown into the face of a smoker who wants to give up smoking.***

*As tobacco became a major commodity during the 16th century, European traders advertised their product with wooden effigies which became known as "cigar store Indians". The statues suggested the exotic and faraway source of tobacco, which by 1600 had been introduced to Sweden, Russia, Turkey, Egypt, Persia, India, and China. So popular was the leaf in the New World that it was legal tender for the payment of wages, debts, and taxes in many colonies.*

# CHAPTER 1

# TOBACCO'S PAST

The Huron Indians of North America have passed down a legend concerning the origin of tobacco. According to the legend, there was once a great famine, when all the lands were barren. The Great Spirit sent a naked girl to restore the land and save his people. Where she touched the ground with her right hand, potatoes grew and the earth was fertile. Where she touched the ground with her left hand, corn sprang forth, bringing green to the lands and filling all stomachs. Finally, the naked messenger of the Great Spirit sat down; and from her place of rest grew tobacco.

There are two interpretations of this tale. One is that tobacco was a gift like corn and potatoes and was meant to provide food for the mind of man. The second is that since tobacco was given by the seat of the messenger, it was intended as a message (or possibly a curse) that the gifts of food were not without their price.

Whichever interpretation is correct, the use of tobacco was firmly established among North American Indians by the time Columbus arrived. The early explorers were amazed to discover the Indians putting little rolls of dried leaves into their mouths and then setting them afire. Some Indians carried pipes in which they burned the same leaves so they could "drink" the smoke. It was also apparent that tobacco was essential to many religious and social rituals and that it was a habit not given up easily.

## *Tobacco Takes Hold*

In the 16th century two British sea captains persuaded three Indians to return with them to London. The Indians brought substantial supplies of tobacco to sustain them through their voyage and stay. That trip may have marked the birth of the Indians' revenge against the white invaders of their land, for along the way some crew members tried inhaling the tobacco smoke. Many enjoyed it and soon found it hard to stop. To supply their own needs for tobacco, the explorers of the 16th and 17th centuries kept fields around the Horn of Africa, in Europe, and in the Americas. Magellan's crew smoked tobacco and left seeds in the Philippines and other ports of call. The Dutch brought tobacco to the Hottentots. The Portuguese brought it to the Polynesians. Soon, wherever sailors went—in Asia, Africa, even Australia—tobacco was waiting.

By the beginning of the 17th century, tobacco cultivation previously only in small plots had been extended to plantations throughout the world. Wherever it was grown, the inhabitants also tried smoking it, thus expanding its use even further. Smoking spread almost like a contagious

*A Mayan Indian smokes a string-tied cigar.*

disease, from a few individuals to entire populations.

Early tobacco users quickly learned what the Indians had long known: once you started, you couldn't stop without severe discomfort and a powerful urge to resume the habit. Furthermore, these needs could only be satisfied by tobacco, which had to be used in certain ways. It did not have to be smoked. It could be chewed or ground to a powder and inhaled as "snuff". Simply eating the raw plant, however, did not provide relief or pleasure, and no other substance seemed to be an adequate substitute.

## *Opposition to the Use of Tobacco*

As tobacco was introduced to one empire after another, a similar pattern of response occurred. What happened in England, under the rule of King James I, near the turn of the 16th century, was typical. King James greatly opposed the use of tobacco. He decried its use as unhealthy and immoral,

***Sir Walter Raleigh's servant discovers too late that he is extinguishing not a fire but Sir Walter's pipe. Raleigh, the British explorer and adventurer, made tobacco fashionable in England during the late 1500s.***

and he urged its banishment. However, even among the royal court tobacco had its dedicated followers: men like Sir Walter Raleigh made tobacco use both fashionable and a mark of distinction.

Attempts to restrict supplies only increased the value of tobacco and soon it was worth its weight in silver. Finally, in one of the earliest recorded attempts at prohibition by economics, King James increased the tobacco duty tax by 4,000%. The only real consequence was to help stimulate a flourishing smuggling trade.

In the end, the economic issues conquered the rulers and not the plant. Seeing that the people would pay nearly any price for tobacco, monopolies were started so that the government could benefit from the desires of its people.

***A South American Indian medicine man summons up the magical powers of tobacco to drive out the spirit which causes disease. Many primitive societies considered tobacco therapeutic.***

Taxation policies were more carefully implemented and the government itself was soon dependent on the trading of tobacco.

This general pattern of disapproval, failed attempts at prohibition, and economic gain by taxation was repeated in Italy, France, Russia, Prussia, and then the United States. As governments became convinced of the dangers of tobacco use, taxes were raised, providing the dual benefit that the conscience of the government was cleansed while its income was enhanced as people contined to smoke. In Britain in 1983 smokers spent a total of £6,200 million on tobacco, of which 60 per cent went to the government in tax. Similarly, in the United States today smokers spend about $16 billion per year (see Figure 1).

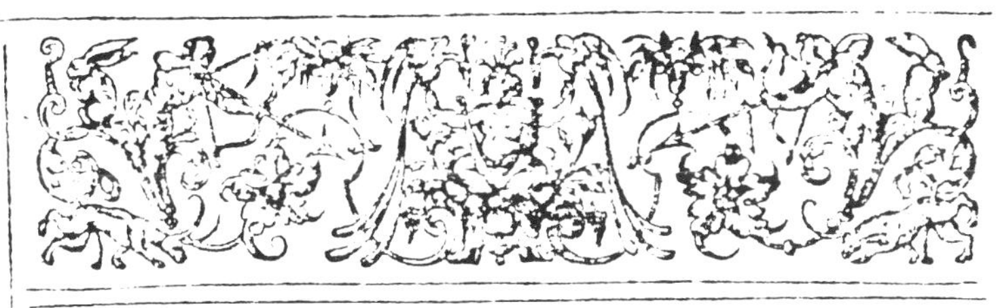

A COVNTERBLASTE
TO TOBACCO.

Hat the manifold abuſes of this vile cuſtome of *Tobacco* taking, may the better be eſpied, it is fit, that firſt you enter into conſideration both of the firſt originall thereof, and likewiſe of the reaſons of the firſt entry thereof into this Countrey. For certainely as ſuch cuſtomes, that haue their firſt inſtitution either from a godly, neceſſary, or honourable ground, and are firſt brought in, by the meanes of ſome worthy, vertuous, and great Perſonage, are euer, and moſt iuſtly, holden in great and reuerent eſtimation and account, by all wiſe, vertuous, and temperate ſpirits: So ſhould it by the contrary, iuſtly bring a great diſgrace into that ſort of cu-

***The opening page of an anti-smoking tract by King James I of England.***

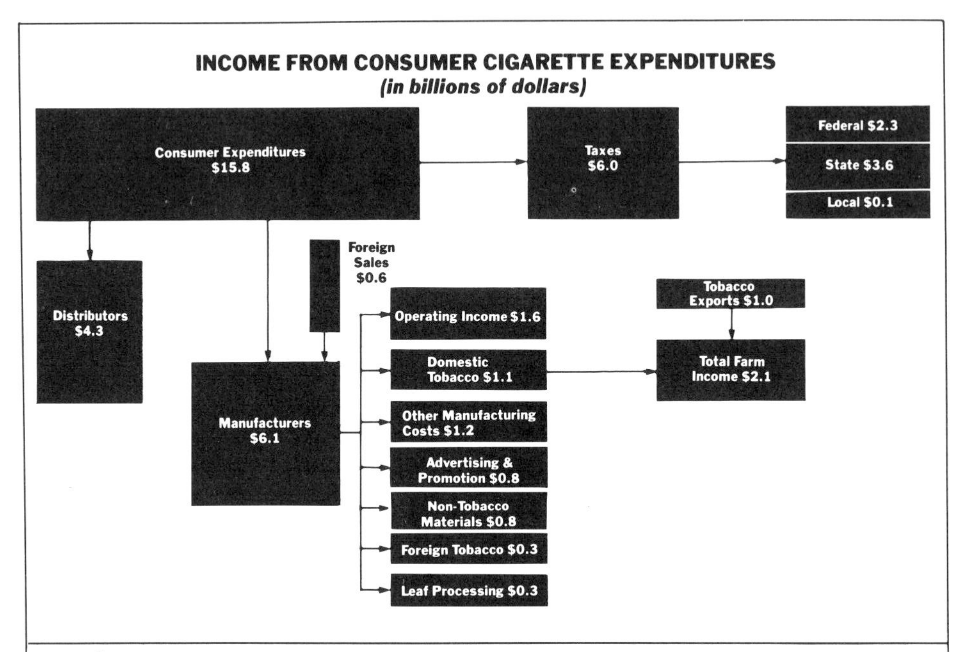


**Figure 1.** ***The chart shows how the $16 billion tobacco industry was distributed in the United States in 1977. Six companies accounted for nearly all of the cigarettes sold that year.***

As a "revenge of the Indians" the spread of tobacco makes the introduction of syphilis into the new world pale by comparison when the enormous toll in death and disease is considered. Tobacco addiction has shared with prohibition the fact that it was a consequence of a behaviour never to be eliminated, either by law, taxation, or papal mandate. Perhaps the most dramatic description of the Indians' revenge was given by "the tobacco fiend" at the court of Lucifer in an 18th century epic:

> *Thus do I take revenge in full upon the Spaniards for all their cruelty to the Indians; since by acquainting their conquerors with the use of tobacco I have done them greater injury than even the King of Spain through his agents ever did his victims; for it is both more honourable and more natural to die by a pike thrust or a cannon ball than from the ignoble effects of poisonous tobacco.*

***An idol dressed as a god holds a cigarette. Tobacco was widely used in religious rituals by the Aztec Indians of Central Mexico.***

***In 1612 John Rolfe planted the first commercial tobacco crop in Jamestown, Virginia. His curing system greatly contributed to Virginia's later prosperity. Here, in 1962, actors portray Rolfe and his wife as part of a Jamestown celebration marking 350 years of the tobacco industry.***

*Tobacco crops in West Virginia await harvesting. Between 1968 and 1981 America's cigarette market grew from $10.1 billion to $18 billion. Although the number of smokers declined during that period, those who did smoke were smoking more than ever before.*

# CHAPTER 2

# WHAT IS IN TOBACCO SMOKE?

What is the nature of this substance which has caused so much controversy and is of such concern to governments, religious bodies, and science? The tobacco plant is a member of the vegetable family *Solanaceae.* The plant was named *Nicotiana tabacum* in honour of the French ambassador to Portugal in the 1850s, Jean Nicot, who believed the plant had medicinal value and encouraged its cultivation. One chemical constituent of tobacco is nicotine. When cultivated with various chemical fertilizers and insecticides, processed into cigarettes, and finally burned, many other physical constituents result.

## *Primary Tobacco Smoke Constituents*

Tobacco smoke contains thousands of elements. Most are delivered in such minute amounts that they are not usually considered in discussions of the medical effects of cigarette smoking. In fact, there are so many that it will take years of research to discover which constituents are harmful. Three of undisputed importance, however, are tar, carbon monoxide, and nicotine.

***Tar*** is defined, rather arbitrarily, as the total particulate matter (TPM), minus water and nicotine, which is trapped by the Cambridge filter used in smoke collection machines. Persons who have used ventilated cigarette holders (for example, MD4) in an effort to give up smoking have probably noticed the accumulation in the filter of a thick, black material reminiscent of road tar. Tar, not present in unburned tobacco, is a product of organic matter being burned in the presence of air and water at a sufficiently high temperature. Tobacco products such as snuff and chewing tobacco do not deliver tar.

Official figures for tar delivery of cigarettes which are published and, in some countries printed on cigarette packets, do not reflect tar contained in the tobacco or even in the smoke. These estimates reflect the amount collected from the standard cigarette-smoking machines. The levels may be useful for cigarette comparisons, but are otherwise misleading to people who think that their intake of tar is mainly determined by their brand of cigarettes. One study showed that very low-tar cigarettes with official yields of only a few milligrams delivered 15–20 milligrams when smoked the way a person might actually smoke them.

Tar is one of the major health hazards in cigarette smoking. It causes a variety of types of cancer in laboratory animals. Also, the minute separate particles fill the tiny air holes in the lungs (the alveoli) and contribute to respiratory problems such as emphysema. In the light of these facts many cigarette manufacturers have reduced the tar yields in their cigarettes in an effort to provide "safer" cigarettes. Unfortunately, tar is important to the taste of the cigarette and the satisfaction derived from smoking. Thus, when many people smoke low-tar cigarettes, to get maximum enjoyment they inhale so deeply that they defeat the purpose of this type of cigarette. It is ironic that cigarettes engineered to deliver low-tar yields when smoked by machines deliver higher yields when smoked by people.

***Carbon monoxide*** (CO) is a gas that results when materials are burned. Carbon monoxide production is increased by restricting the oxygen supply, as is the case in a

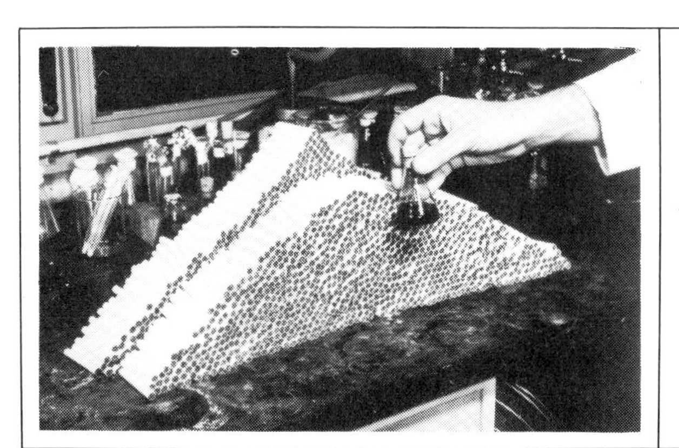

***Two thousand cigarettes, the number stacked on the table, produce the amount of tar contained in the flask. Tar is a major reason why cigarette smoking is dangerous.***

cigarette. Carbon monoxide is also produced by internal combustion engines (as in cars) and even by gas cookers and heaters. Like carbon dioxide ($CO_2$), which also results from burning, carbon monoxide easily passes from the alveoli of the lungs into the blood stream. There it combines with haemoglobin to form carboxyhaemoglobin (COHb). Haemoglobin is that portion of the blood which normally carries carbon dioxide out of the body ($CO_2$ is produced by normal metabolic processes) and oxygen back into the body. When the haemoglobin is all bound up by either carbon monoxide or carbon dioxide, a shortage of oxygen may result.

A critical difference between carbon monoxide and carbon dioxide is that the former binds much more tightly to haemoglobin and is very slow to be removed. Thus the blood can accumulate rather high levels of carbon monoxide and slowly starve the body of oxygen. When the cardiac system detects insufficient levels of oxygen, the heart may begin to flutter and operate inefficiently. In extreme cases a heart attack may result.

Figure 2 shows carbon monoxide levels in cigarette smokers who smoked different numbers of cigarettes. These values were collected by having a group of smokers smoke one of their usual brand of cigarettes every 30 minutes until they had smoked either 5 or 10 cigarettes. The figure shows that each cigarette causes a brief boost in the CO level which lasts for a few minutes and then declines until the next cigarette is smoked.

However, each cigarette adds slightly to the overall

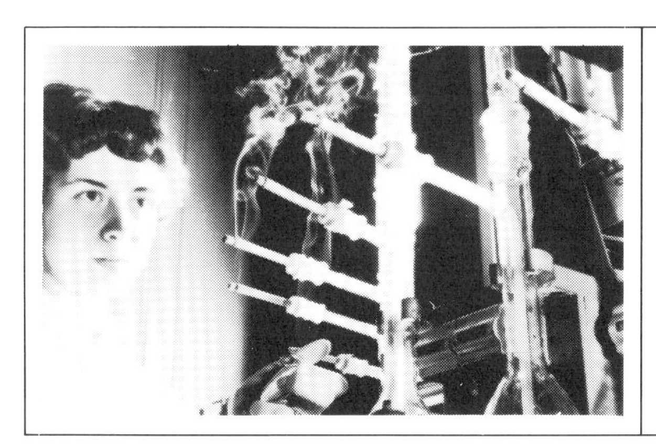

***A laboratory technician analyzes the effectiveness of filters in capturing tar, nicotine, and other elements of cigarette smoke. While simulation of the many variables remains difficult, the use of computer models in recent years has substantially aided the research effort.***

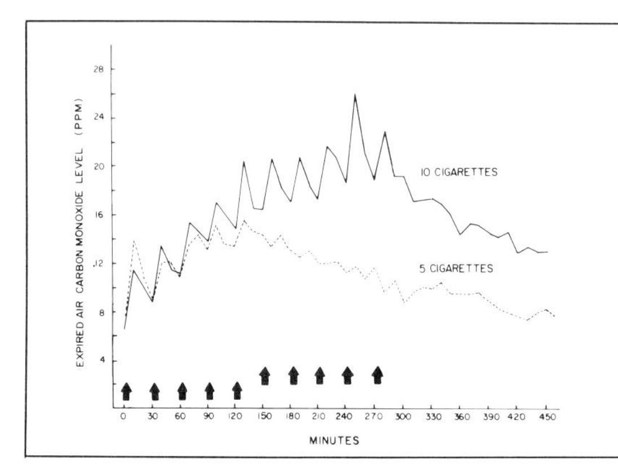


***Figure 2.** Average carbon monoxide (CO) levels are shown for eight volunteers who smoked one cigarette every 30 minutes as indicated by the arrows. On one day they smoked five cigarettes (dotted line), and on another day they smoked ten cigarettes (solid line).*

level. When people smoke normally, their CO levels are lowest in the morning and level off at their highest values by midday. The typical twenty-per-day smoker achieves levels averaging between 25 and 35 parts per million. However, even these "average" smokers may hit short-term levels of greater than 100 parts per million. Firefighters are now routinely checked with portable CO analyzing machines while combating fires. If their levels exceed 150 parts per million they may be relieved and given oxygen, since even these generally healthy people run a risk of heart attacks.

***Nicotine*** is a drug that occurs naturally in the leaves of *Nicotiana tabacum.* It is generally thought of as a stimulant since it provokes many nerve cells in the brain and heightens arousal. However, its effects are so complex that no simple label is completely accurate. For instance, by stimulating certain nerves in the spinal cord (Renshaw cells), nicotine relaxes many of the muscles of the body and can even depress knee reflexes. Its effects also vary depending on how much is smoked. For example, nerve cells that are stimulated by the nicotine from a few cigarettes may be depressed by smoking more cigarettes.

Nicotine closely resembles one of the substances that occur naturally in the body (acetylcholine), and the body has an efficient system to break nicotine down (detoxification) and eliminate it in the urine (excretion). In fact, when a given dose of nicotine is ingested, for instance by smoking, about one-half is removed from the blood stream within 15 to 30 minutes. Figure 3 shows the pattern of nicotine accumulation and elimination when cigarettes are smoked. Note that these patterns are similar to those observed for carbon monoxide (see Figure 2).

Another feature of nicotine is that it is well absorbed through the mucosae or the very thin skin of the nose or mouth which is dense with capillaries. This is why chewing tobacco and taking snuff are effective ways to ingest nicotine. In the form of cigarette smoke, nicotine transfers directly from the alveoli of the lungs into the arterial blood stream and rushes directly to the brain. It requires less than 10 seconds for inhaled nicotine to reach the brain, so that even though the quantities are small, the effects may be strong. This is even quicker than giving nicotine intravenously, since the venous blood supply must first pass through the heart, then into the arterial stream of the lungs and, finally, to the brain.

Repeated exposure to nicotine, when it is smoked, results in very rapid tolerance or diminished effect. That is, during the day, as cigarettes are smoked, the smoker get less and less of a psychological and physical effect—even though toxins are building up in the body. In fact, much of this tolerance is lost overnight. As a result, cigarette smokers often report that their first cigarette of the day "tastes the best" or may even make them lightheaded if smoked too quickly. As the day wears on and more cigarettes are smoked, people often smoke more out of habit or to avoid discomfort than for pleasure.

Research to date on the hazards of cigarette smoking suggests that nicotine is not as hazardous as the tar and

***Tobacco often achieves the status of currency wherever it is hard to obtain. This Australian aborigine, puffing on his crab-claw pipe in 1949, would have performed almost any task for payment in tobacco.***

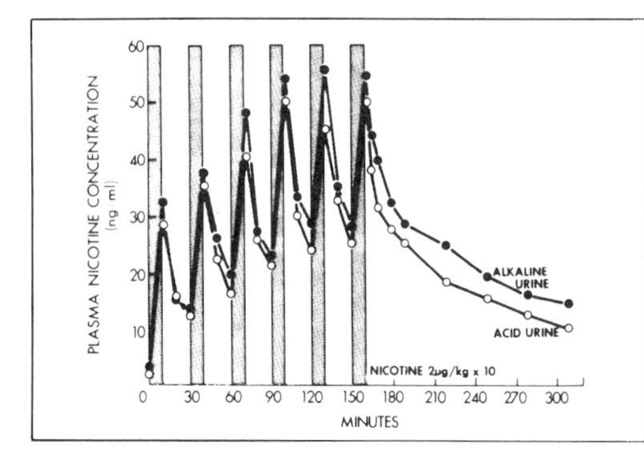


**Figure 3.** *Five volunteers were given injections of nicotine. Average nicotine levels in plasma are shown by the circles. Acidification of the urine resulted in more rapid nicotine excretion and lower plasma levels of nicotine, showing the body's capacity to eliminate nicotine levels in the bloodstream.*

**Figure 4.** *Automatic cigarette smoking machines continue to make a great contribution to smoking research. Technicians can modify various aspects of puffing, such as volume and duration of intake, at the push of a button. The smoke is drawn into the machine and stored in special containers.*

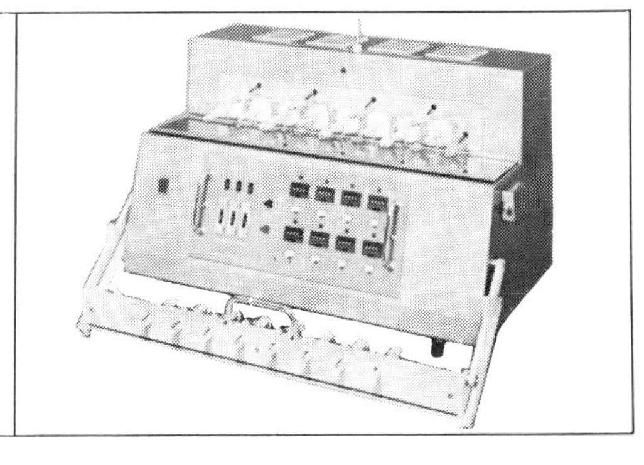

*A hospital employee in San Diego takes part in an experiment which sought to determine how quickly the lungs heal once a smoker has quit the habit. Recent research indicates that lung performance improves markedly soon after the termination of smoking.*

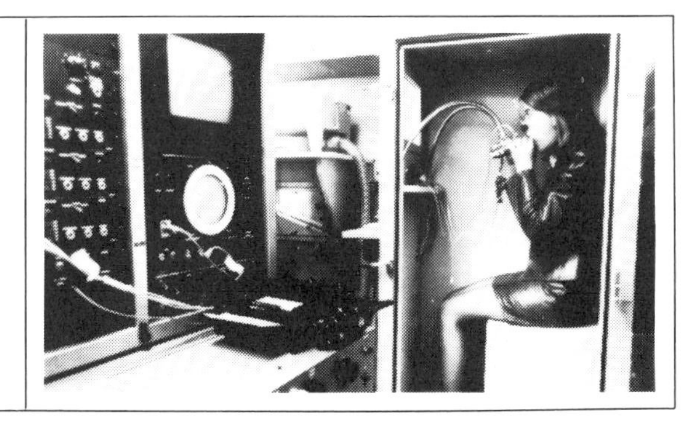

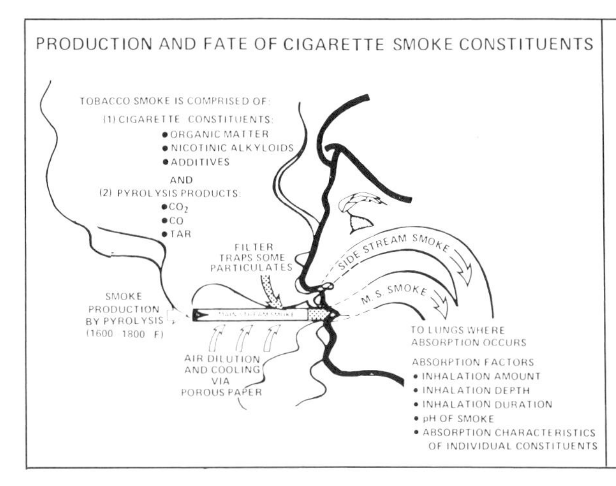


**Figure 5.** ***The diagram illustrates what happens when a cigarette is smoked. The burning of the cigarette creates various substances (i.e. carbon dioxide, carbon monoxide, and tar) in addition to the cigarette's constituents prior to burning.***

carbon monoxide in cigarette smoke. Its role is more insidious, since people ingest these other substances (tar and CO) as a by-product of their efforts to obtain nicotine.

### *Other Constituents of Tobacco Smoke*

Cigarette smoke is made up of both a gas phase and a particulate phase. Together they include more than 4,000 substances. Automatic cigarette-puffing machines have been devised to collect and to study smoke. Figure 4 shows such a machine, which might be used by research laboratories or cigarette manufacturers. The smoke is separated into the gas and solid (particulate) phases by passing it through a filter pad (Cambridge filter), which traps particles larger than one micrometre and collects the rest (gas phase) in a storage tank. The machines are calibrated to smoke the cigarettes the way a typical smoker might smoke them.

Figure 5 shows what happens during a puff. The unburned cigarette is comprised of many organic (tobacco leaves, paper products, sugars, nicotine) and inorganic (water, radioactive elements, metals) materials. The tip of the burning cone in the centre of the cigarette reaches a temperature of nearly 1,093 degrees C. (2,000 degrees F.) during a puff. This tiny blast furnace results in a miniature chemical plant, which uses the hundreds of available materials to produce many more. In fact, some of the most important parts of tobacco smoke (including tar and carbon monoxide) are not even present in an unburned cigarette, but are produced when a puff is taken and the cigarette burns.

Study of the smoke is made even more complicated since there are both sidestream and mainstream smoke which must be separately collected and studied. The mainstream smoke is collected from the stream of air passing through the centre of the cigarette. It is filtered by the tobacco itself and perhaps further by a filter. It is also diluted by air passing through the paper (most modern cigarettes also have tiny ventilation holes which further dilute the smoke).

Sidestream smoke is that which escapes from the tip of the cigarette. It is not filtered by the cigarette and results from a slightly cooler burning process at the edge of the burning cone. Since the tobacco is therefore burned less completely, the sidestream smoke has more particulate (unburned material) in it.

### *Cigarette Engineering*

The above process is complicated even further by the engineering efforts of the tobacco manufacturers. They specifically construct cigarettes in ways to control a wide range of factors: keeping the cigarette burning between puffs, reducing spoilage of the tobacco, altering the flavour of the smoke, and controlling the amounts of substances (tar and nicotine) measured by government agencies.

The porosity of cigarette paper is specifically controlled to regulate the amount of air that passes through and dilutes the smoke. Porosity also affects how rapidly the cigarette

***Machinery representing the state of the art in cigarette manufacture at R.J. Reynolds Tobacco Company, in 1961. Increasingly sophisticated production methods gave America's tobacco industry 61% of the world's tobacco export market at that time.***

burns. Phosphates are added to the paper to ensure steady and even burning.

Several kinds of additives are present in the tobacco itself. One type of additive is called a *humectant.* Humectants are chemicals that help retain the moisture (humidity) of the tobacco. This is important in how the tobacco burns. Humectants also affect the taste and temperature of the smoke. The most commonly used humectants are glycerol, D-sorbitol, and diethylene glycol. Humectants comprise a few per cent of the total weight of the tobacco.

Another type of additive is called a *casing agent.* This helps blend the tobacco and hold it together. It also affects the flavour of the smoke and how quickly the tobacco burns. Most commonly used casing agents include sugars, syrups, licorice, and balsams. The amount of casing agents used ranges from about 5% of the total weight of the tobacco in cigarette tobacco to about 30% of the weight of pipe tobacco.

Specific *flavouring agents* are also added to the tobacco to control the characteristic taste of a cigarette. These include fruit extracts, menthol oils, spices, coca, aromatic materials, and synthetic additives. Flavour is also controlled by curing processes and, of course, the type of tobacco itself.

A variety of other substances are added at various stages of tobacco processing to retard spoilage and prevent tobacco larvae from developing into worms. In addition, metals such as nickel and potassium are taken up from the soil, as are pesticides and fertilizers used in tobacco farming. There are also radioactive elements such as potassium-40, lead-210, and radium-226, which result from fallout and the natural background.

***Designed to attract young smokers, flavoured Nova cigarettes went on the market in Japan in 1983. Manufacturers have continued to imrove the naturally harsh taste of tobacco with many different flavouring additives.***

***A detail from a temple carving shows a Mayan Indian priest smoking as part of a religious ceremony. Mayan sacred writings indicate that smoking was widespread in Central America long before Europeans arrived in the 16th century.***

# CHAPTER 3

# TOBACCO AND HEALTH

Tobacco has been used for its social and medicinal effects. In *Don Juan*, Sganarelle touts tobacco as "the passion of all proper people, and he who lives without tobacco has nothing else to live for. Not only does it refresh and cleanse men's brains, but it guides their souls in the ways of virtue, and by it one learns to be a man of honour."

Tobacco has been used at various times to treat headaches, asthma, gout, labour pains, and even cancer. Its diverse behavioural effects have also been noted. It was used by monks to suppress sexual drive. The philosopher Kant, on the other hand, regarded tobacco as a way of provoking sexual excitement. King James I wrote, "Being taken when they go to bed, it (tobacco) makes one sleep soundly, and yet being taken when a man is sleepy and drowsy, it will, as they say, awake his brain and quicken his understanding." (Generally an opponent of tobacco use, King James nonetheless made many accurate observations of the diverse effects of tobacco.)

Not all observers regarded tobacco so benignly. One of the strongest condemnations also came from King James I in his "Counterblaste to Tobacco". He called tobacco use a custom "lothesome to the eye, hatefull to the nose, harmfull to the brain, dangerous to the lungs, and in the black stinking fumes thereof, neerest resembling the horrible stinking smoke of the pit that is bottomeless." It is interesting to point out that some have attributed King James' anti-smoking stance to his personal dislike of Sir Walter Raleigh, who strongly promoted the use of tobacco.

## *How Many Teenagers Smoke*

Children become aware of smoking when they are very young, and over a quarter have had at least a few puffs of a cigarette by 10 years of age. Recruitment to regular smoking takes place mainly in the early teens, and from about 5% at age 11, the number increases sharply so that by 16 about 25% are regular smokers. Although boys used to greatly outnumber girls among smokers, the past decade has seen a trend for smoking to increase in girls to match boys. Now in most countries there is approximate equality between the sexes, and in some there are hints that girls may be overtaking boys.

Many factors influence children to start smoking, but among the most important is parental example. Children whose parents smoke are much more likely themselves to take up the habit. About 25% of 10- to 12-year-old smokers report having had their first cigarette with their parents and being given cigarettes by their parents. Peer example is also a key influence. The majority of young smokers' friends smoke, and smoking in the teenage years usually take place in the company of other young people.

Although the level of smoking in young people is

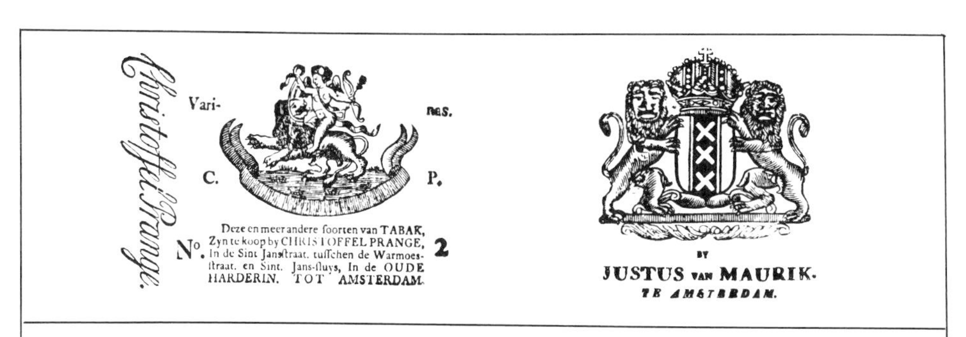


***Ornate Dutch tobacco wrappings demonstrate the fact that even in the 18th century, packaging of tobacco was an important aspect of marketing and sales.***

alarmingly high there has been some decline in recent years, parelleling trends in older people. Alongside this reduction there have been accompanying changes in attitudes about smoking. Between 1975 and 1980 the percentage of high school students in the U.S.A. who viewed smoking as a "great risk" increased from 51% to 64%, and surveys in the early 1980s indicate that more than 70% "disapprove" of smoking twenty or more cigarettes per day, and more than 40% believe that smoking should be prohibited from certain public places.

It is also clear that the impact of the addictive nature of cigarettes has not been fully realised since less than 1% of teenagers who smoke say that they will "definitely" be smoking in five years and only 13% say that they will "probably" be smoking in five years. The sad truth is that the

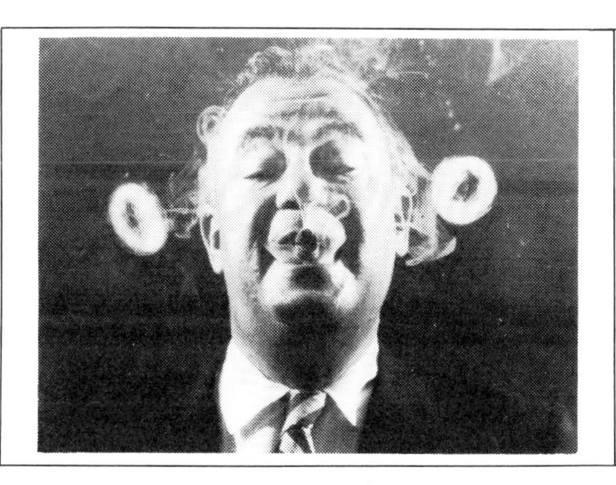

***An American entertainer demonstrates his ability to blow perfect smoke rings. Like other addicts, smokers develop their own set of rituals.***

***This seven-year-old is only pretending to smoke during a tobacco auction in Georgia, but the desire to imitate adult behaviour is one of the main reasons that many young people begin smoking.***

overwhelming majority will still be smoking five years later and for many years after.

### *Health Consequences of Cigarette Smoking*

Since the first reports of the ill effects of cigarette smoking began to appear in the 1950s a vast literature has accumulated incriminating smoking as the leading preventable cause of premature death in the Western world. It is estimated that smoking kills 100,000 people every year in the U.K. and over 300,000 per year in the U.S. These numbers completely dwarf the numbers of deaths caused by other external factors such as alcohol, road accidents, suicides, etc. The following quotation from the most recent report of the Royal College of Physicians helps give some appreciation of the massive scale of the problem:

"Among 1,000 young male adults in England and Wales

*Mother: 'Augustus, you naughty boy, you've been smoking. Do you feel very bad, dear?'*
*Augustus: 'Thankyou—I'm only dying.'*

who smoke cigarettes on average about

1 will be murdered
6 will be killed on the roads
250 will be killed before their time by tobacco.''

***Cancer*** is the second leading cause of death in the developed world. Smoking causes at least 30% of all cancer deaths. The most common form of cancer caused by smoking is lung cancer, which accounts for about 30,000 deaths in smokers every year in the United Kingdom. The rate of lung cancer in women is rising steeply in all Western countries, paralleling the increase in female smoking. Smoking is also the major causal factor in cancers of the larynx, mouth, and oesophagus. It is a contributing factor (co-factor) in the development of cancers of the bladder, pancreas, and kidney. Pipe and cigar smoking has also been implicated in cancers of the lung, larynx, mouth and oesophagus.

The evidence is strong enough to conclude that cigarette smoking is not only associated with these illnesses, but that it is a major cause of cancer. A cigarette smoker is 10 times more likely to die of lung cancer than a nonsmoker. This risk increases with the number of cigarettes smoked, but is partially reversible by stopping. Death due to cancer in

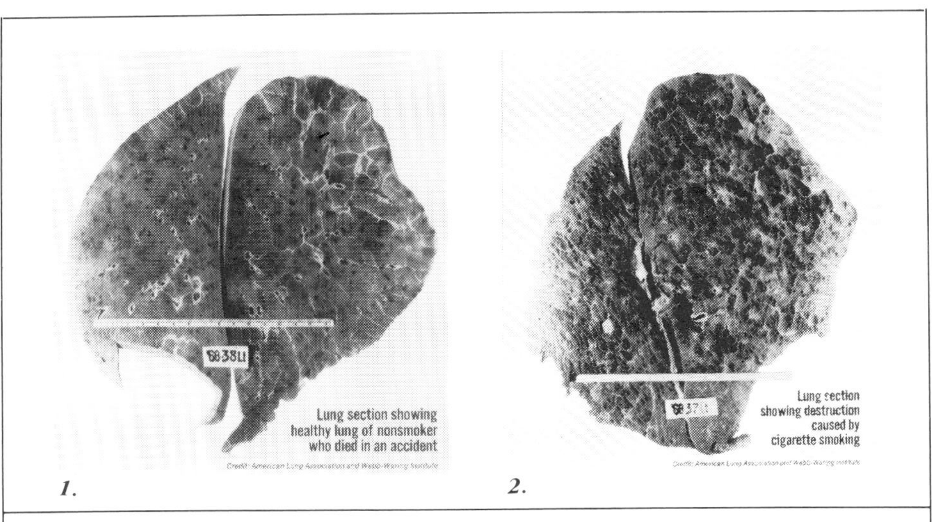


***The above posters show (1.) a healthy lung and (2.) a diseased lung.***

former smokers, who stopped 15 years ago or longer, is only slightly greater than that of nonsmokers. Since 1950 the lung cancer rate has increased more than 5% a year among American women, who started smoking in large numbers after World War II.

Available statistics suggest that cancer of the lung may soon overtake breast cancer as the major cause of cancer death for women. Preliminary studies suggest an increased risk of lung cancer in nonsmoking wives of smoking husbands, implicating sidestream smoke as a cancer risk factor. Controlled animal studies have shown exposure to cigarette smoke increases various kinds of cancer, even when the animals are otherwise well-fed and cared for.

***Chronic obstructive lung disease*** includes the diseases of chronic bronchitis and emphysema. It causes distressing and disabling breathlessness, and accounts for an enormous number of days lost from work every year. After prolonged ill health it often proves fatal. About 20,000 people die from this condition in England and Wales every year, and although other factors such as social class and air pollution are partly responsible, the dominant causative factor is now cigarette smoking. The death rate from chronic bronchitis is six times as great in smokers as in non-smokers.

Apart from these major respiratory illnesses in which the air passages of the lungs are narrowed and damaged and much of the lung tissue is destroyed, it has also been shown that "normal" smokers have far less pulmonary function than nonsmokers; that is, the ability of the respiratory system to provide oxygen to the blood is impaired. This loss of function is, in part, due to direct damage of several sorts. For instance, elastin, a major structural protein of the lung, is adversely affected by cigarette smoking. Also, evidence suggests that the small airways of the lung may be adversely affected in healthy nonsmokers if they are exposed to the cigarette smoke of others. Lung performance on standard breathing tests (spirometric evaluation) improves soon after one stops smoking. However, pulmonary diseases are a progressive condition and are best treated by early detection and termination of smoking.

***Coronary heart disease*** (CHD) is one of the leading causes of death in developed countries, and leads to 180,000 deaths each year in the U.K. and 750,000 per year in the U.S. Cigarette smoking is one of the three major risk factors for coronary heart disease, the other two being high blood pressure and high levels of cholesterol in the blood. Smokers are at a higher level of risk of heart disease than nonsmokers, particulary in early middle age, and the more one smokes the greater the risk. Cigarette smoking is thought to cause about 36,000 deaths from CHD each year in the U.K.

Stopping smoking very quickly reduces the risks of coronary heart disease, and one year after a person has given up smoking their risk is reduced to the same level as that of the nonsmoker.

Smoking is a major factor in many other diseases. It has important effects in pregnancy, and increases the risk of spontaneous abortion and of deaths of babies around the time of birth. There is clear association between smoking and peptic ulcers in the stomach. It leads to diseases of arteries elsewhere than in the heart. These and other conditions have been extensively reviewed in the reports of the U.K. Royal College of Physicians and of the U.S. Surgeon General (see Further Reading). Smoking is responsible for 15 to 20 per cent of all deaths in Western industrialised countries, and of all environmental causes of death is the major one that could be avoided.

## *The Benefits of Smoking*

Scientific studies of the effects of cigarette smoking reveal that many of the early claims about the benefits of tobacco, as well as the above described dangers, were not groundless. Some previously suggested benefits were based on valid observations. Smoking is a convenient way for people to help regulate their mood and feelings, and it has some benefit in the control of body weight.

When possible therapeutic benefits of a drug are considered we must address the issues of "safety" and "efficacy". The drug must be proved safe in the form in which it will be marketed, and the drug must be proved effective in scientific tests. These are key factors used by the U.S. Food and Drug Administration (FDA) when new drugs or food additives are considered for approval.

The nicotine delivered by cigarette smoking clearly has

some therapeutic effects. On the other hand, the overwhelming evidence that cigarette smoke also causes a wide range of damaging effects would prevent tobacco from being approved for therapeutic purposes. However, to understand why so many people smoke, why treatment of smokers is difficult, and why most people who stop smoking soon resume the habit, it is important to understand some of the benefits gained by smoking.

As a means of *self-treatment* tobacco is very convenient. It is also legal, relatively inexpensive, and readily available. It may be consumed nearly anywhere with minimal social stigma, the amount (dose) can be precisely controlled, and it is in a convenient delivery system. In the form of tobacco

***A movie still from 1933 reinforces an attitude toward smoking later successfully exploited by such advertisements as "Come to Marlboro Country". Until the recent increase in public awareness of the dangers involved, smoking was often associated with manliness in the popular imagination.***

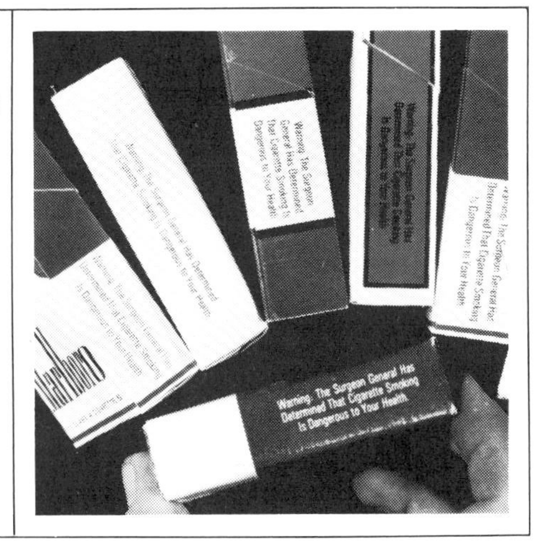


***When the Surgeon General's warning first appeared on cigarette packets in 1964, smoking declined immediately throughout the United States. Even stronger warnings on packets took effect in 1985.***

smoke, nicotine can be delivered to the blood stream within seconds of smoke inhalation. The nicotine dose has extremely rapid and controllable effects. In fact early researchers dubbed this system "finger-tip control of dose", to emphasize the puff by puff precision with which a smoker can regulate his or her nicotine dose level.

Once in the blood stream, nicotine has immediate effects on many hormones. Two, in particular, have been studied in detail: adrenaline and noradrenaline. Adrenaline is released into the bloodstream when people are anxious, stressed, or bored. Noradrenaline is released during certain kinds of heightened arousal caused by excitement, exercise, antidepressant drugs, sex, many drugs of abuse, and nicotine. There is some evidence that cigarette smokers can use nicotine to adjust their levels of noradrenaline and thus self-regulate their own moods and emotional states. It should come as no surprise that two of the most commonly given reasons for smoking are for "stimulation" and "reduction of stress".

Studies show that nicotine has some therapeutic effects probably related to these biochemical effects. In one study the effects of nicotine on aggression were measured. Two research subjects were seated in separate rooms in which they worked at a computer game to earn money. One subject would periodically lose points (and money). This loss was thought, by the subject, to be due to the deliberate actions of the other. When this happened the person who lost the money could either subtract points from the other person or blast the subject with a loud noise. Before the test the subject was given a cigarette to smoke. The cigarettes

***In Japan 75% of all adult males smoke. The government-owned industry sold over 300 billion cigarettes in 1982, up 2.7 billion from 1981.***

delivered different doses of nicotine. The higher the dose of nicotine that the "angry subject" smoked, the less likely he was to punish the other subject.

Other studies have verified that nicotine can indeed reduce anxiety and make people more tolerant of stressful events and distractions. Together, these studies support claims by many smokers that cigarettes help them to deal more effectively with the daily stresses in life.

Tests have also shown that in certain kinds of tasks involving speed and reaction time, nicotine can improve performance. It also helps on tasks involving *vigilance and concentration*, such as watching a computer screen. The enhancing effects of nicotine on performance are strongest when cigarette smokers have been deprived of cigarettes for several hours. However, the enhancing effects also occur when nonsmokers are given small doses of nicotine in tablet form (buffered so that it would be absorbed through the stomach).

Many people claim that they smoke as a means of *weight control*, and there is evidence that nicotine and smoking does facilitate this in at least three ways. Firstly, nicotine decreases the efficiency with which the body extracts energy from food. Thus more food is eliminated without being converted to fat or muscle. Secondly, nicotine reduces the appetite for foods containing simple carbohydrates (sweets). Lastly, smoking reduces the eating that often occurs as a response to stress. As a result of these effects, cigarette smokers tend to weigh less than nonsmokers.

Roughly one-third of the people who stop smoking gain

***Tobacco-chewing members of the New York Yankees satisfy their nicotine habit while avoiding any intake of tar. Tar is produced only when the organic matter in a cigarette burns in the presence of air and water.***

weight. Anti-smoking zealots point out that this is not so bad since two-thirds of the people who stop smoking do not gain weight. But they miss an important point. That is that weight gain is a possible undesirable side effect to stopping smoking. For any other type of therapy, a side effect that occurred in one-third of those treated would be considered worthy of serious attention.

It should be apparent that the effects of smoking are many and varied, and that some short-term effects may be beneficial. In the long run, smoking is hazardous to health and shortens the expected life span. Understanding the benefits, however, can provide insights as to why some people smoke. More importantly, it should be clear that removing the benefits of smoking may lead to side effects which must be addressed. Otherwise treatment will not be successful.

***A Barnum & Bailey's Circus trainer poses in 1913 with "Old Joe", the camel whose likeness still appears on packets of Camel cigarettes.***

*An American Cancer Society poster points out that smoking during pregnancy can have adverse effects on the unborn child. Associated problems include retarded growth, spontaneous abortion, and malformation. Until adolescence, children of mothers who smoke remain several months behind children of nonsmokers in reading and maths.*

# CHAPTER 4

# SMOKING, WOMEN, AND PREGNANCY

Cigarette smoking among women is a more recent phenomenon than it is for men. Whereas tobacco consumption was already very high in British men at the start of World War II, British women only took to smoking in large numbers from this time.

Smoking in British men began to decline in the early 1960s as influential reports, such as those from the Royal College of Physicians, linking smoking and cancer, were widely publicized (see Figure 6 for U.S. comparison). These reports were thought to be a critical factor in reducing the incidence of smoking in men, particularly those who were better educated and had higher incomes.

Smoking rates in women continued to rise as women became more active in social roles and fields of work that had been dominated by men. Smoking in women peaked in the early 1970s, and now, as in men, is firmly on a declining trend.

Early in the twentieth century cigarette advertising was aimed only at men. But as the 1920s brought women new

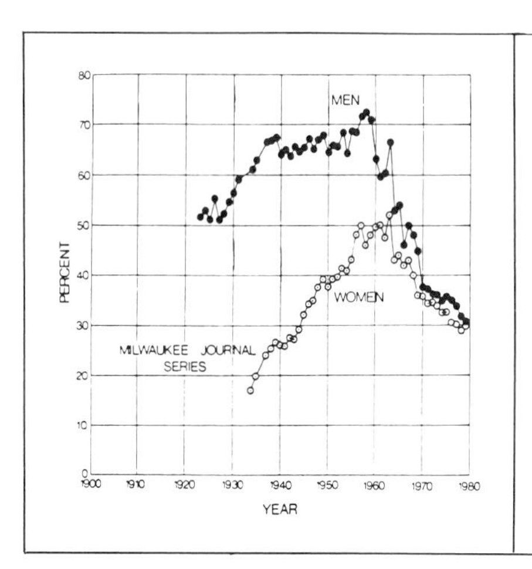


**Figure 6.** ***These data were collected in annual surveys of residents in the Milwaukee area by the*** **Milwaukee Journal.** ***The graph shows the percentage of men and women who reported smoking cigarettes during each of the years indicated on the horizontal axis. The substantial decline in the incidence of smoking after 1960 is evidence of increasing public awareness of the dangers of smoking.***

freedoms, American tobacco companies quickly recognized the potential of the female market. Chesterfield advertisements in 1926 had young women pleading to young men smokers "Blow some my way". By using the slogan "Reach for a Lucky instead of a sweet" American Tobacco was able to sell cigarettes to women as an alleged means of losing weight. In a series of massive advertising campaigns, the company appealed directly to women, using testimonials from well-known women such as Amelia Earhart, the famous flyer, and film star Jean Harlow.

In the 1970s cigarette manufacturers marketed brands that identified smoking with female liberation. Philip Morris launched Virginia Slims with the slogan "You've come a long way, baby", and by 1976 it had become *the* women's cigarette. In Britain the slogan was changed to "We've come a long, long way" in response to objections to the sexist "Baby" slogan, but otherwise the theme of the woman who discovers her emancipation through her cigarettes remained the same. The early 1980s saw tennis player Martina Navratilova promoting British American Tobacco's products on worldwide television as her dress at Wimbledon sported the Kim logo.

## *The Effects of Smoking on Women*

It is a myth that women are immune to the adverse effects of cigarette smoking. In both men and women, the prevalence of all major types of chronic respiratory diseases (bronchitis, asthma, impaired breathing) is directly related to the level of smoking. Similarly, as women have begun to smoke more, their incidence of lung cancer has shown a steady rise. Additionally, women who take birth control pills are at greater risk of cardiovascular disease if they smoke cigarettes.

There is even evidence that cigarette smoking affects some aspects of the sexuality of women. Smoking more than half a packet of cigarettes per day is associated with a higher incidence of infertility. Irregular menstrual cycles are more prevalent among women who smoke cigarettes. The number of years of potential fertility is also reduced. Menopause occurs earlier among women who smoke.

### *The Effects of Smoking on Offspring*

The effects of cigarette smoking on pregnancy, birth weight, and infant health have been studied extensively. Figure 7 shows the effects of cigarette smoking on birth weight and several other aspects of pregnancy. Babies born to women who smoke are an average of 200 grams (about 7 ounces) lighter than babies born to nonsmokers. This is important since birth weight is an excellent predictor of infant health. It appears that this retarded growth is caused by hypoxia or decreased oxygen available to the foetus. This is partly due to the carbon monoxide delivered by smoke inhalation.

Another effect of cigarette smoking during pregnancy is to increase the likelihood of spontaneous abortions. In fact,

***Until the United States government banned cigarette advertising on radio and television, tobacco companies traditionally sponsored shows and used guest stars to endorse their product.***

the risk is almost double for women who smoke. Smoking also increases the risk of congenital malformations. This, like so many other effects of smoking, is directly related to the amount of smoking. Levels of smoking are also associated with a variety of other complications during pregnancy and labour. These include increased risk of bleeding and premature rupture of membranes. Finally, there is a clear relationship between smoking during pregnancy and the occurrence of the sudden infant death syndrome (SIDS).

Babies born to cigarette smokers develop more slowly throughout childhood than babies born to nonsmokers. They are more likely to have neurological (brain function) disorders, psychological abnormalities, and lower intelligence scores. Until adolescence, children of mothers who smoke 10 or more cigarettes per day remain about three to five months behind children of nonsmokers in reading, mathematics, and general ability scores. Cigarette smoking during pregnancy is also a significant risk factor for hyperkinesis in children.

The mechanisms that underlie the harmful effects of

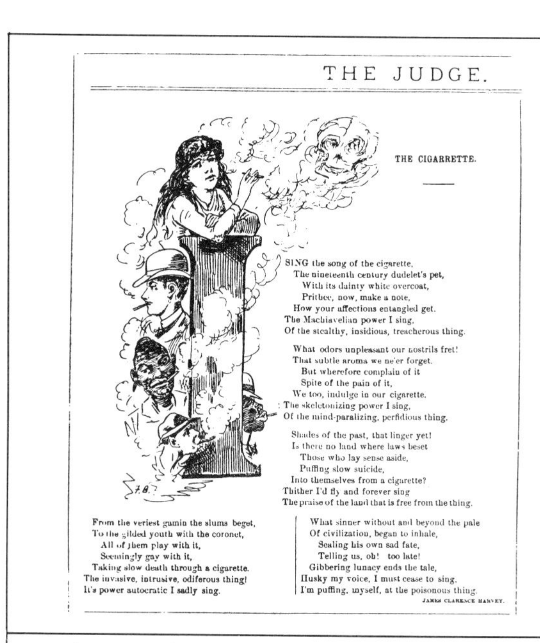

THE JUDGE.

THE CIGARETTE.

SING the song of the cigarette,
The nineteenth century dudelet's pet,
With its dainty white overcoat,
Prithee, now, make a note,
How your affections entangled get.
The Machiavelian power I sing,
Of the stealthy, insidious, treacherous thing.

What odors unpleasant our nostrils fret!
That subtle aroma we ne'er forget.
But wherefore complain of it
Spite of the pain of it,
We too, indulge in our cigarette.
The skeletonizing power I sing,
Of the mind-paralizing, perfidious thing.

Shades of the past, that linger yet!
Is there no land where laws beset
Those who lay sense aside,
Puffing slow suicide,
Into themselves from a cigarette?
Thither I'd fly and forever sing
The praise of the land that is free from the thing.

From the veriest gamin the slums beget,
To the gilded youth with the coronet,
All of them play with it,
Seemingly gay with it,
Taking slow death through a cigarette.
The invasive, intrusive, odiferous thing!
It's power autocratic I sadly sing.

What sinner without and beyond the pale
Of civilization, began to inhale,
Sealing his own sad fate,
Telling us, oh! too late!
Gibbering lunacy ends the tale,
Husky my voice, I must cease to sing,
I'm puffing, myself, at the poisonous thing.

JAMES CLARENCE HARVEY.

***A turn-of-the-century poem by James Harvey satirizes the smoking habit.***

***A cigarette advertisement from the 1920s aims squarely at image-conscious female smokers.***

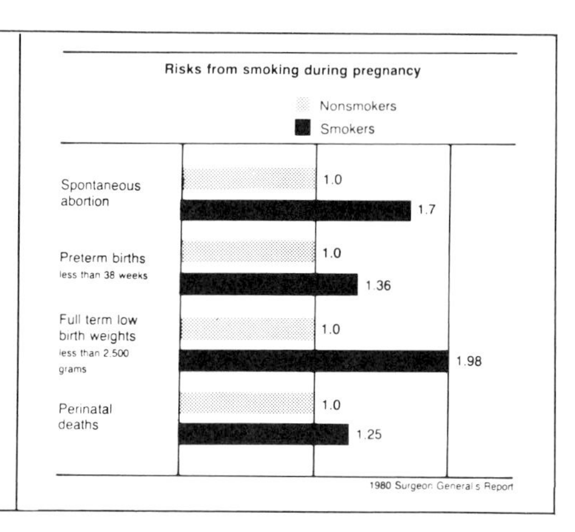


**Figure 7.** ***This chart shows the number of various kinds of pregnancy problems for women who smoke cigarettes for every one of such problems found in women who do not smoke.***

smoking on the foetus and infants may be hypoxia (oxygen starvation) due to the CO carried by the smoke into the lungs. Nicotine, which crosses the placenta, may raise the blood pressure and thus induce slowing of the heart. Tar also crosses the placental barrier but detrimental effects have not been clearly proved to date. These data make it clear that cigarette smoking is a risk factor for both mother and infant. While smoking does not ensure damage, it is an avoidable risk factor. Since much of the risk is to the child-to-be, the responsibility of the mother is great indeed.

***A North Carolina child rides a sled of tobacco leaves to a curing barn in 1949 during the first weeks of the yearly auction markets.***

*A technician at the Louisiana State University School of Medicine lights up a cigarette for a baboon as part of a study investigating hardening of the arteries caused by smoking. Scientists have used animals as experimental subjects in smoking-related tests since the 1950s, when serious research into the medical effects of smoking first began.*

# CHAPTER 5

# THE SCIENTIFIC STUDY OF CIGARETTE SMOKING BEHAVIOUR

Until a few years ago there had been little effort to study the behaviour of cigarette smokers. This is in sharp contrast to the enormous expenditure in money and research hours spent studying the effects of smoking on health. Even the amount of money aimed at trying to convince or help people to give up smoking far surpassed that spent on trying to understand the behaviour itself.

## *Resistance to Study of Cigarette Smoking Behaviour*

Intensive research into why people smoke did not begin until the late 1970s. There were scientific, political, and practical reasons for this. Until recently, cigarette smoking was held to be normal voluntary behaviour that was a source of pleasure. Advertising campaigns have been careful to develop and perpetuate this view of smoking. Cigarette smoking was presented as a means of building a particular image, perhaps one of sophistication.

In the 1920s and 1930s cigarettes were endorsed by famous athletes and film stars, who attributed part of their own success to smoking a particular brand. In the U.S. the recent successful images have been those of the rugged Marlboro Man, the dedicated and independent Camel smoker, and the confident, liberated Virginia Slim woman. Another technique commonly used by cigarette advertisers is to associate smoking with the beauties of nature and healthy outdoor living.

Since the scientific study of behaviour is usually reserved for actions considered deviant, pathological, or related to mental health or drug problems, cigarette smoking, which was seen as normal and voluntary behaviour, received little study. The view of cigarette smoking as a voluntary pleasurable act was most prevalent from the early 20th century until recently. Thus, it was reasoned, smokers simply needed to be informed that smoking was harmful to their health; then they could weigh the risks and benefits in a rational manner, and presumably, would stop. The other side of the coin is that when a person failed to stop smoking, it was assumed that he or she did not really understand that smoking was harmful, or that the person had a self-destructive personality.

This logic led to anti-smoking campaigns which showed lung cancer operations. For some people, informing them that smoking was harmful was sufficient to cause them to stop. However, as King James learned in the early 17th

*Every 20 seconds for five years until September 1977, the typically rugged man portrayed in this Chicago cigarette advertisement took a puff and blew perfect smoke rings. Tobacco companies spend vast amounts of money on marketing every year.*

century, for most people such campaigns were of little lasting value.

The political and economic reasons that delayed the study of cigarette smoking behaviour are intertwined. Tobacco represents a vast industry and gives employment to thousands of people. In the United Kingdom tobacco is the Chancellor of the Exchequer's third biggest source of consumer revenue, and tobacco companies have been given large subsidies to foster production and employment. Over a period when the government gave just over £2 million for the anti-smoking activities of the Health Education Council, over £29 million was given to the tobacco industry in grants. In the United States tobacco is an important part of the rural economy, and the government subsidizes the tobacco industry in much the same way that it subsidizes many other facets of the agricultural industry. Such subsidies are critical to many state and local economies for which tobacco is a major source of revenue.

***In 1959, leaders of the Seventh Day Adventist Church, working in conjunction with physicians, devised a programme designed to help smokers give up smoking without the use of drugs. The programme, presented in five sessions, concentrates on the physical, mental, social, and spiritual implications of smoking.***

From a practical standpoint, studying the behaviour of cigarette smokers also represented a considerable challenge. Social, cultural, pharmacologic, and behavioural variables are involved in smoking. The behaviour itself is complex. It was not clear which aspects of cigarette smoking were important to measure. The number of cigarettes smoked tells little about how they were smoked or how much cigarette smoke was taken in by the lungs. Number of puffs or volume of smoke inhaled are not easily automated. It was not even clear how to control the dose of smoke given.

The fundamental questions of what aspects of cigarette smoking to measure and how to measure them remained stumbling blocks to cigarette smoking research until recently.

***When the American government declared copper an essential commodity during World War II, the makers of* Lucky Strike *cigarettes were suddenly deprived of the metal needed for printing the brand logo. They promoted their loss as a contribution to the war effort, claiming that their sacrifice had released enough copper to build many light tanks.***

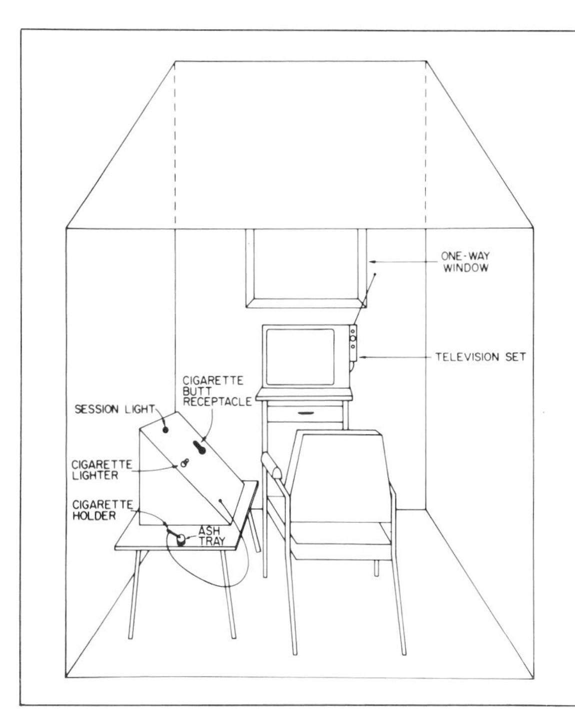


**Figure 8.** ***The instruments in this cigarette smoking test room are wired to a computer in an adjacent room so that every puff on every cigarette is automatically measured.***

## *Scientific Laboratories and Devices to Study Cigarette Smoking*

Before studies of cigarette smoking could proceed, laboratories and devices had to be developed to study the behaviour. Laboratories were developed to study smoking at a macro level (where larger aspects of smoking are studied, such as the effect of a cold on the number of cigarettes smoked) and a micro level (which might consider what happens during individual puffs). For the most controlled level of study, test rooms were developed in which volunteers could smoke cigarettes while computers recorded their behaviour. One such test room is shown in Figure 8. In this context it was possible to study carefully the behaviour of cigarette smokers and to discover what factors controlled it.

The cigarette smokers who served as subjects in these studies did not feel that they smoked differently from normal while in the laboratories. However, it was important to prove this, in other words to validate the system. To do this, portable puff-monitoring systems were developed. These systems take the scientist one step away from the laboratory where events are more controlled, and one step closer to the natural environment of the smoker. These allowed smoking behaviour to be monitored by people as they went about their daily living in their usual environments.

Another type of laboratory setting makes less use of automated devices but permits rigorous observation of cigarette-smoking behaviour. This is the residential research laboratory where volunteers live in the company of other volunteers for periods of a few days to several months. In such a setting, people are free to smoke as they choose, but they obtain each cigarette either from cigarette dispensing machines or from research staff. Patterns of smoking are thus observed 24 hours a day.

Each of these settings is uniquely arranged to measure some aspect of smoking behaviour. The strategy of using several systems permits a variety of checks (points of validation) to be made. It also provides a much more complete picture of smoking.

### *Descriptive Studies of Cigarette Smoking Behaviour*

The first step for scientists using these laboratories and devices was to measure carefully all that happens when

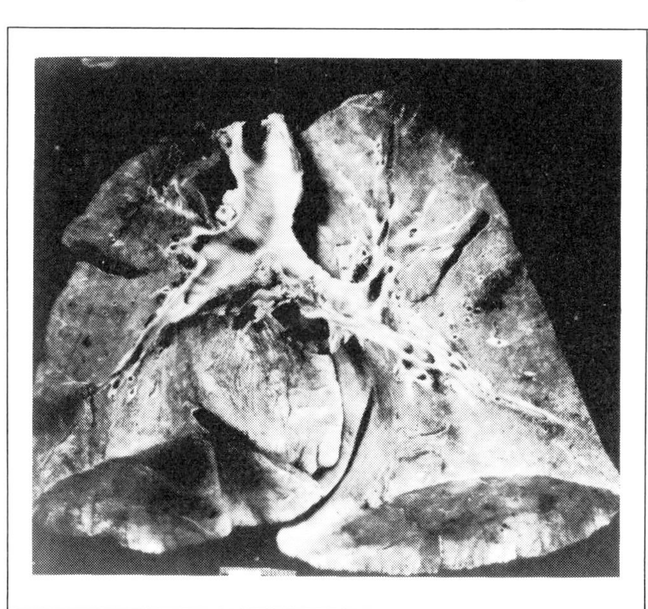

***The healthy lungs of a non-smoker reveal tissues which can be damaged by exposure to smoke exhaled by others.***

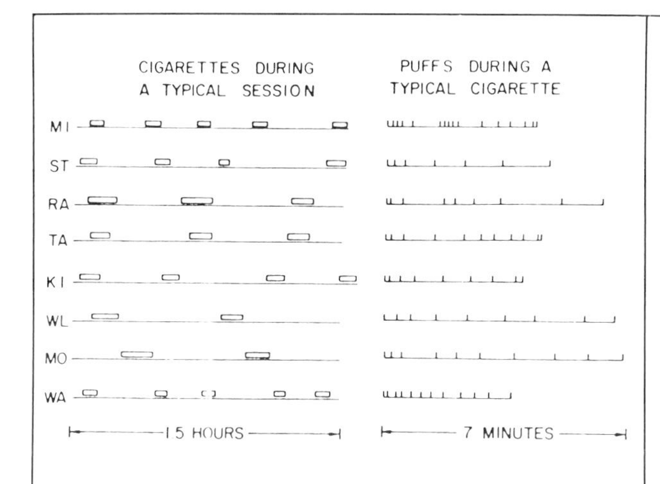


**Figure 9.** ***Typical patterns of smoking cigarettes (left side) and puffing on each cigarette (right side) are shown for each of eight volunteers. In the course of the experiment, smoking emerged as an orderly routine in which addicts compensated for deprivation by smoking more intensely when a cigarette was finally given.***

people smoke cigarettes. The first studies recorded the times when people smoke. It was learned that cigarette smoking was not a random event, but rather an orderly routine. Figure 9 shows smoking patterns obtained in one laboratory study. Similar patterns were recorded whether by staff observations or by portable puff-monitoring devices. Furthermore, the relationship between number of cigarettes smoked and CO levels was similar whether or not puff-monitoring systems were used. Since each system gave the same result, it was likely that the results were valid.

Another set of studies showed that puffing and inhaling were also very orderly. As the cigarette grew shorter, puffs tended to become smaller in duration and volume. It also appeared that when strong cigarettes were smoked, many people took smaller puffs but diluted them by taking in greater volumes of air when they inhaled the smoke. Depriving smokers of cigarettes resulted in their smoking more cigarettes and taking larger puffs when they were again permitted to smoke. The findings suggested that cigarette smoking was controlled to a large extent by a drug contained in the tobacco smoke.

### *The Effect of Different Features of Tobacco on Smoking*

When drugs of abuse are studied, the dose, or the amount of drug taken, is a critical factor. The general finding is that larger doses are taken less often than smaller doses, resulting in relatively stable amounts of the drug being taken. This effect is termed "compensation" or "dose-regulation" and occurs in many forms of drug use. For instance people tend to drink beer in much greater volume than whisky. With regard to cigarette smoking this issue of compensation has both theoretical and practical implications.

Theoretically, it is important to discover if people regulate cigarette smoke the same way they regulate intake of drugs. If they do, then it is important to discover precisely what they are regulating: is it the smoke itself or the nicotine?

Practically, it is important because one way of treating smokers is to have them gradually decrease the number of cigarettes that they smoke. People who cannot stop smoking

***During the 1930s many celebrities, including star athletes, were contracted by tobacco companies to attribute part of their success to smoking. Despite the many accomplishments of Mr. Gehrig, the fact remains that fitness and smoking are rarely compatible.***

are also encouraged to smoke cigarettes delivering less tar and nicotine on the assumption that they are safer. But smokers who compensate for changes in the strength of their cigarettes by smoking more of them or inhaling more deeply will defeat their purpose.

One study used an automatic puff-monitoring system to measure smoking behaviour when subjects were required to smoke through ventilated cigarette holders. (These holders were sold to help people to give up smoking by gradually cutting back their dose.) The main finding of the study was that the smaller the dose permitted by the filter, the more puffs were taken. These smokers did not smoke many more cigarettes, but with the extra puffs they were able to take in nearly as much smoke. This was proved by the fact that their CO levels were similar regardless of the holder used.

Two other approaches to manipulating dose were changing the number of puffs permitted per cigarette and cutting cigarettes into various lengths. The general finding of these

*"But, Dad, it's not like when you were a boy. The smoke in these cigarettes is pulled through tiny, scientific, mentholated air pores, where thirty-five thousand filters trap the nicotine and tars, and protect the throat from harmful irritation."*

approaches is the same as that in the ventilated holder study. Namely, people compensate for changes in dose and end up taking in about the same amount. In these studies more cigarettes were smoked when the cigarettes were shorter, or when fewer puffs were permitted per cigarette.

These studies showed that people regulate tobacco smoke intake quite well, but was it the smoke or the nicotine contained in the smoke that they were regulating? Until recently uncertainty existed because several studies had shown that if special cigarettes were given in which the smoke was the same in all respects except nicotine level, people did not appear to regulate the number of cigarettes they smoked as well as would be expected. The results of the ventilated study offered a clue as to what might be happening. Namely, smokers might regulate in subtle ways, like number of puffs or the depth of inhalation.

The way to address the issue was to measure the amount of nicotine actually taken into the body. Several such studies were done in the 1980s. These studies showed that the amount of nicotine in the blood had little to do with the amount of nicotine in the cigarette. It appeared that the smokers regulated their behaviour so that they obtained a certain nicotine level regardless of the dose in the cigarette.

These studies uncovered another point of similarity

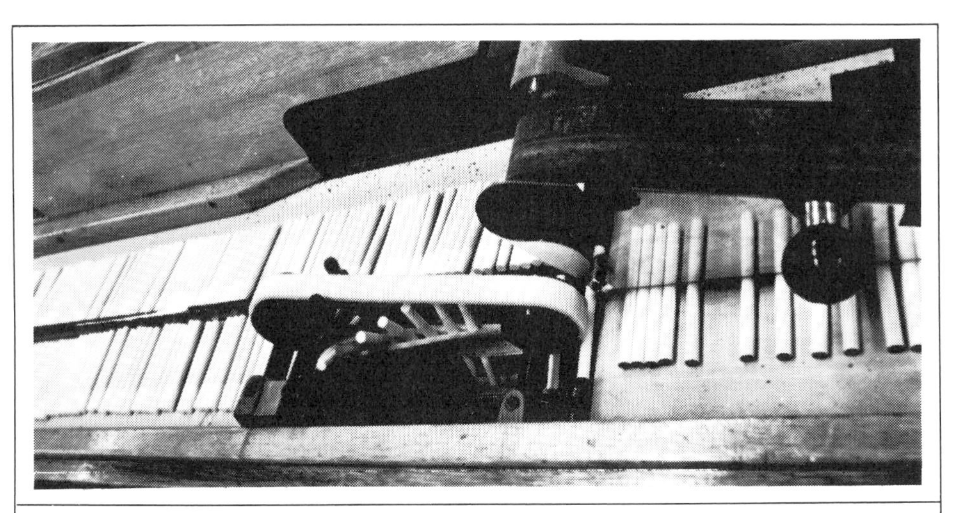

*A conveyor belt in a cigarette factory, capable of carrying up to 500 billion cigarettes each year.*

between the use of tobacco and other drugs of abuse. The dose of the substance is a critical factor in controlling the pattern of drug intake. The practical implication of this finding is that instructing people to switch to lower tar and nicotine brands of cigarettes may result in a false sense that they have lowered their health risk. The evidence indicates, in fact, that smokers of lower tar and nicotine cigarettes take in nearly as much of the substances contained in smoke as smokers of stronger cigarettes. This also helps explain the poor results of such approaches to giving up smoking as ventilated holders and switching brands.

### *Other Drugs and Smoking*

Drugs of abuse generally have a variety of behavioural effects. It is important to learn of these effects on cigarette smoking

***Blacks predominated in process work in the tobacco industry throughout the 19th century. industry in the U.S. through the 19th century. This picture was taken in a Virginia tobacco factory.***

for both theoretical and practical reasons. Theoretically, discovering how certain drugs affect cigarette smoking behaviour can help us to understand the nature of the behaviour itself. From a practical standpoint it is important to learn if drugs can be used to help treat smoking, or if taking drugs will hinder treatment efforts.

One of the first drugs to be carefully studied for its effects on smoking was alcohol. It is commonly claimed by smokers that they smoke more when they drink. The laboratory studies proved this claim objectively and went a few steps further.

First, the studies showed that it was not just a social phenomenon. It even occurred when people drank in isolation. Second, they did not simply light up more often. The measures of puffs taken and CO levels showed that people actually inhaled more. Third, it was not simply due to the person's expectation: it happened when the flavour of the beverage was masked; the stonger the dose of alcohol the more they smoked. It was dependent, however, on past experience of drinking and smoking, because the effect was very weak in light social drinkers, and alcohol may actually decrease smoking in people who hardly drink at all. Sedatives (pentobarbital) and opioid narcotics (methadone and

***In this 18th-century caricature entitled "One of the Swinish Multitude", artist Richard Newton displays an attitude towards smoking shared by many nonsmokers today.***

morphine) also increased cigarette smoking in abusers of these drugs.

One drug of particular interest was *d*-amphetamine. Amphetamines have been prescribed to help people stop smoking on the theory that cigarettes are stimulants and that substituting one for another should reduce smoking. However, the theory is oversimplified (smoking can both stimulate and relax), and the treatment has not worked. Figure 10 shows that when given *d*-amphetamine people actually smoked more. This study measured several factors and showed that people smoked more cigarettes, took more puffs, achieved higher levels of smoke intake (CO), smoked the cigarettes down further, and felt better while smoking.

The findings regarding *d*-amphetamine suggest an interesting theory. Namely, drugs that produce positive feelings (euphoria) may increase cigarette smoking. Certain drugs such as amphetamine appear to be somewhat more univer-

***Actress Brooke Shields lends her support to a campaign by the National Council on Alcoholism. Researchers have discovered that experimental subjects smoke more intensely and more frequently when drinking alcoholic beverages.***

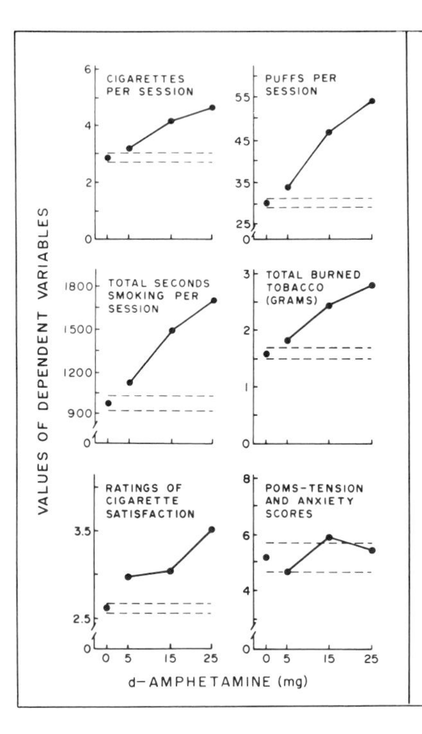


**Figure 10.** ***Eight volunteers were tested during 90-minute sessions during which several aspects of cigarette smoking (shown on the vertical axis) were measured. Two hours before each session they were given a capsule that contained either a single dose of amphetamine or a placebo (shown on the horizontal axis.) The zero on the horizontal axis stands for the placebo, and thus on each graph the point above the zero and the area within the dotted lines represents the normal, drugless, state. Points falling outside the dotted lines indicate that the effect of the amphetamine was significantly different from that of the placebo. The POMS scale (right bottom panel) shows that the effect was not due to nervousness.***

sal euphoriants and may affect most smokers in the same way. Other drugs such as alcohol may only be euphoriants in certain individuals and may increase smoking only in these people. Caffeine has relatively little effect on smoking. This is probably because at the same time that caffeine provides some stimulation and euphoria, it can also make people more nervous. These findings on the effects of psychoactive drugs have important treatment implications. One is that many of the drugs that an individual commonly uses while smoking cigarettes should be avoided while he or she is trying to give up smoking cigarettes.

The findings of the above studies led to a hypothesis that drugs that blocked the effects of nicotine in the body might alter cigarette smoking regardless of their effect on mood. Therefore, it was predicted that treatment with nicotine not in tobacco form would reduce cigarette smoking. This is because, even though nicotine has some euphoriant effects, it would be like giving a person much stronger cigarettes and he or she should compensate by smoking less. In several studies volunteers were given nicotine either in an intravenous form or in the form of a chewing gum. The results were consistent with the hypothesis; that is, people smoked less.

A drug that diminishes the effects of nicotine should increase cigarette smoking as the smoker attempts to compensate. A study was done in which an antihypertensive medicine, mecamylamine, was given in capsule form. According to the theory, if the dose of nicotine is suddenly reduced because it has been blocked, the person should smoke more. The results were consistent with the hypothesis; that is, people smoked more. Theoretically, if such a blocker were given continuously, the person would ultimately stop smoking since they would not get any pleasure out of smoking. Preliminary testing proved this to be true. These results suggest that such medicine might ultimately help people to stop smoking if repeated on a daily basis.

THE HISTORY AND MYSTERY OF TOBACCO.

TO what extent active stimulants are necessary for the health of the body and the development of the intellect, affords a subject of speculation which, it seems, will never be brought to a satisfactory conclusion. Speaking without referring to the experience of all ages, we would say that, beyond a sufficiency of wholesome food, nothing more was necessary to sustain the human body in its greatest perfection; yet it is notorious that, from the earliest ages and among all peoples, the custom has prevailed of using a thousand substances, evidently for no other purpose than to give unnatural acceleration to the system; and thus, through the body, add im-

Entered according to Act of Congress, in the year 1855, by Harper and Brothers, in the Clerk's Office of the District Court for the Southern District of New York.

Vol. XI.—No. 61.—A

*A 19th-century tract against smoking raises a question debated to this day: why do human beings take stimulants when such substances are not essential to sustaining life?*

***This Norman Parkinson photograph caused a sensation when it appeared in* Vogue *magazine in 1950. Although cigarette companies had often depicted smoking as a sign of sophistication (especially in advertisements aimed at female consumers), female smokers were still a minority at the time.* Vogue's *editor-in-chief cabled to his foreign associates: SMOKING IN VOGUE SO TOUGH, SO UNFEMININE.***

CHAPTER 6

# EFFECTS OF NICOTINE

The importance of nicotine in cigarette smoking has to do with its effects on the brain. This is not a new revelation. Centuries ago it was recognized that tobacco produced "mind altering" effects, that it was a "food for the brain". We now know that many of these effects are not unique to nicotine but are shared with most other drugs of abuse. To understand why people smoke cigarettes when they know it is harmful to their health, it will help to know what nicotine does to the brain.

## *How Nicotine Acts in the Brain*

To affect the brain, a drug must have physical properties that permit it to pass through the blood brain barrier (a barrier that helps protect the brain from foreign substances). The nicotine molecule has ideal physical features that allow it to penetrate the brain easily. When nicotine is inhaled into the lungs, the arterial blood stream picks it up and carries it to the brain within 10 seconds.

Once in the brain, nicotine molecules work like keys opening locks. The locks are called "receptors" and are located on nerve cells of the brain and connect it to muscle tissues and the various organs of the body. Drugs that act as keys are called "transmitters" since they help to send information from one part of the body to another.

Figure 11 shows a transmitting and receiving station (nerve cell). Receptors located on the receiving end (dendrites) of the nerve cell transmit information by an electric impulse through the axon. At the end of the axon

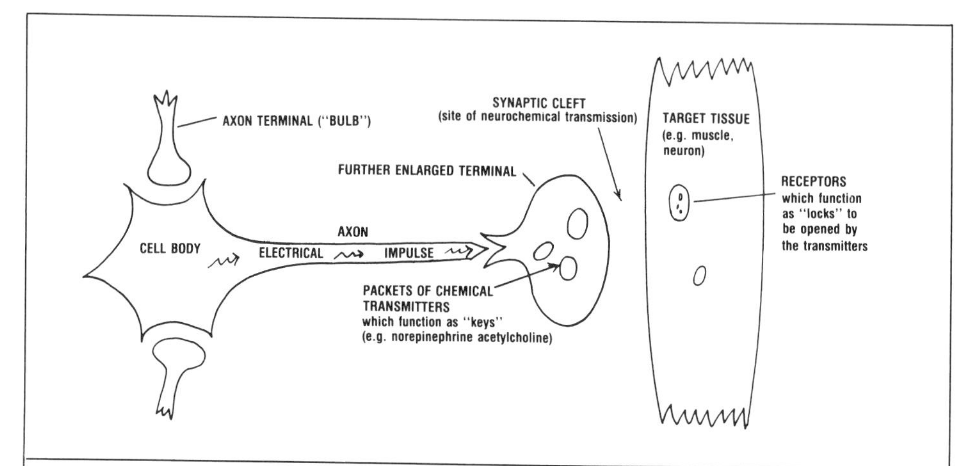


**Figure 11.** *The drawing represents a "typical" neurotransmitting system. The neuron receives a chemical impulse from its dendrites. This is converted to an electrical impulse that passes down the axon to the nerve terminal. There, a small amount of a chemical transmitter is released. The transmitter diffuses across the synaptic cleft where it bonds to a receptor. Most drugs act at this synaptic cleft. For instance, cocaine and nicotine both cause some extra noradrenaline to be released from terminals that use noradrenaline as transmitters.*

terminal a small amount of the body's own chemical transmitter (for instance, adrenaline) is released. The transmitter crosses the gap (synaptic cleft) where it has its effect on the target organ; the target may be another nerve cell which is fired, an organ such as the heart which is speeded up, or a muscle which is stimulated.

Nicotine has complicated actions because the receptors activated by it are located throughout the body. In fact, much of the mapping of the nervous system was done by placing nicotine at various places and charting the effects. This is much like mapping hidden currents in a lake and river system by placing dyes in the water and tracing their course.

Places that were activated by nicotine were called nicotinic receptors. Some of these receptors are located on the adrenal gland. Activation of them results in the release of adrenaline and noradrenaline into the body. These chemicals increase heart rate and blood pressure, heighten arousal, and cause feelings of excitement. Cocaine also results in heightened adrenal activity and has effects on the brain similar to those of nicotine.

A drug that activates a receptor is called an "agonist". Other substances that share some of the critical physical properties of the agonist may substitute for the agonist—much as a master key can substitute for a particular key. For instance, codeine, morphine, and heroin can all substitute for one another. It merely requires adjustment of the dose to get the same effect.

Nicotine acts by substituting for the body's own chemical (acetylcholine) at nicotinic receptors. Lobeline, contained in some over-the-counter aids to help stop smoking, partially substitutes for nicotine. Nicotine in the form of a newly marketed chewing-gum (Nicorette) substitutes very well for nicotine in tobacco smoke. Amphetamine appears to substitute for some of nicotine's effects. Both drugs increase the activity of adrenaline and noradrenaline.

A drug that blocks the effect of an agonist drug is called an antagonist. Some drugs act as antagonists by stimulating other systems (functional antagonists). For instance, giving coffee to a person who is intoxicated by alcohol produces the well-known "wide awake drunk" phenomenon. Other drugs act as antagonists by occupying the receptor without having any effects of their own, thus preventing the agonist drug from working (competitive antagonists). This is like

*A Virginia tobacco farmer and his wife at curing time stoke the fires used to cure the tobacco. This procedure was carried out every two hours for up to two weeks.*

filling a keyhole with wax, thus preventing the key from entering. An antagonist for nicotine is called mecamylamine. Mecamylamine and similar drugs (ganglionic blockers) are used to treat patients with high blood pressure. They are currently being tested as a possible help in stopping smoking since they block the pleasurable effects of nicotine.

### *Physiological Dependence on Tobacco*

Physiological dependence on a drug occurs when nerve cells have adapted so well to the drug that they require the drug for normal functioning. A person is physiologically dependent on a drug when sudden drug abstinence is followed by withdrawal signs. These may be thought of as "rebound" effects. For example, if heart rate is increased by taking a drug, then abstinence may result in decreased heart rate for a few days. Similarly, feelings of contentment that accompany cigarette smoking may be replaced by irritability, and weight loss may be replaced with weight gain. Many other drugs that are abused produce rebound effects. For example, the constipation that normally accompanies narcotic use is replaced by diarrhoea; the muscular relaxation produced by alcohol and sedative drugs is replaced by tremors and even convulsions. When drugs produce extensive

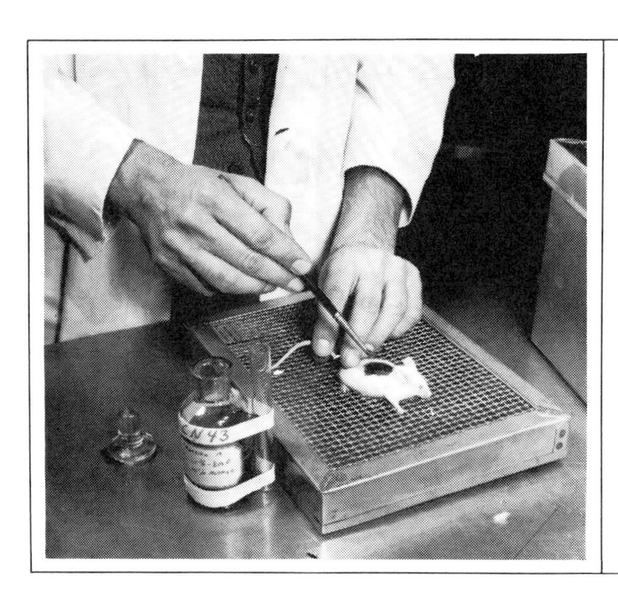

***Mice have been regularly chosen for experimental laboratory tests to determine the various effects of tar and nicotine on living tissue. They were frequently chosen for experiments because they are easy to handle and house.***

and possibly life-threatening rebound syndromes, as is the case with opiates and sedatives, the drug is said to be capable of producing physiologic dependence.

It has not been scientifically proven that nicotine produces a significant withdrawal syndrome. Only about one-third of the people who stop smoking experience physical discomfort, and this is not usually as severe as that associated with drugs such as morphine or alcohol. On the other hand, as with morphine and alcohol, physical dependence on nicotine may only occur in extremely heavy users.

At this time, it appears that addiction to nicotine is more like cocaine dependence than morphine dependence. With all three drugs, behaviour becomes compulsive because of the physical effects of the drug on the brain. The only difference is that abstinence from morphine causes withdrawal, whereas abstinence from cocaine and nicotine may not. These facts have tended to blur the long held separation of physical and psychological dependence.

A common misconception is that drug abuse is the same as physical dependence; it is not. Physical dependence need not occur to produce drug abuse and, if it does occur, it does not ensure that drug abuse will occur. For instance, many people periodically abuse alcohol or narcotics without ever becoming physically dependent. Also, there are

***Neatly bundled cigarette butts are displayed in a Tokyo station by a Japanese citizens group to discourage commuters from dropping litter.***

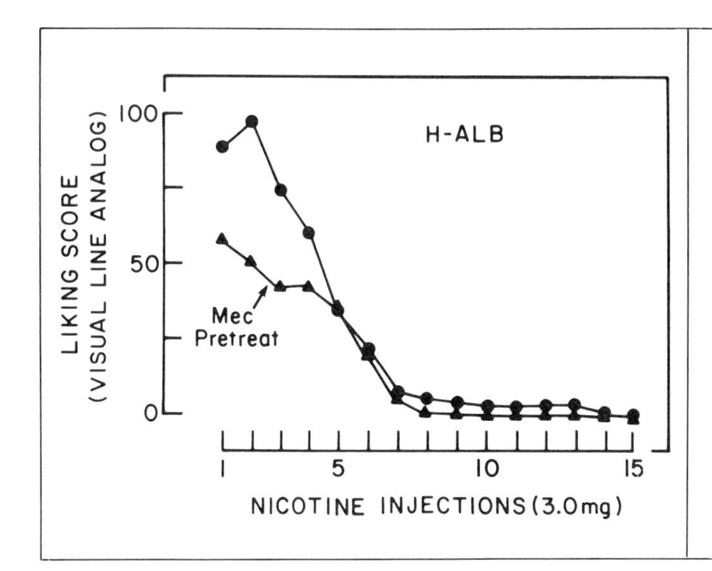


**Figure 12.** ***The decline in the liking score demonstrates the tolerance to nicotine which the body acquires during repeated intake of the drug.***

drugs with an enormously high potential for abuse, such as cocaine and amphetamine, which do not produce a clear-cut syndrome of withdrawal.

On the other side of the coin, a person may become physically dependent on a drug and yet never seek that drug or learn to abuse that drug. This happens frequently when people are given narcotics for pain relief. Only a very small portion of these people abuse the drug or continue to take it when it is no longer medically required.

## *Tobacco Tolerance*

As cigarettes are smoked throughout the day, the effects become smaller and smaller. Often smokers assert that their first cigarettes of the day are the "best tasting", with the rest being smoked out of habit or need. When repeated use of a drug results in smaller and smaller responses, the phenomenon is called "tolerance". In other words, it takes more of the drug to produce the same effect. This explains why the first few cigarettes of the day cause the greatest increase in heart rate and blood pressure, the greatest decrease in the knee reflex, and the strongest psychological effect. A laboratory demonstration of this effect is shown in Figure 12. Repeated injections of nicotine produced progressively smaller psychological effects on the subject.

Tolerance develops when the body becomes more efficient at detoxifying and eliminating the drug (metabolic tolerance). People who have smoked for many years eliminate nicotine from their systems much more quickly than nonsmokers. Tolerance also occurs when the nerve cells become less responsive to nicotine doses.

With nicotine, much of this tolerance is lost overnight and quickly regained when smoking resumes the next day. Finally, tolerance can also occur by means of behavioural adjustments as the person learns to compensate for any disruptive effects of the drug. All of these forms of tolerance occur with nicotine, and all appear to be critical in allowing people to learn quickly to enjoy smoking and to get over sickness that often occurs when it is first tried.

***A 19th-century German cartoon. The introduction of tobacco caused much controversy in Germany, where many officials considered it a threat to civilized behaviour although there was no understanding at that time of the effects of nicotine.***

***A 1953 advertisement for* Lucky Strike *seeks to place the brand firmly in the context of all-American wholesomeness (i.e. outdoor living, barbecue time, the happy couple). Creative advertising encouraging a dangerous addiction continues to this day.***

# CHAPTER 7

# ARE TOBACCO AND NICOTINE ADDICTIVE?

A recurrent theory in the history of the use of tobacco is that its use is compulsive and similar to other forms of drug abuse. In the Americas the inability of the Indians to abstain from tobacco raised problems for the Catholic Church. The Indians insisted on smoking even in church, as they had been accustomed to doing in their own places of worship. In 1575 a church council issued an order forbidding the use of tobacco in churches throughout the whole of Spanish America.

Soon, however, the missionary priests themselves were using tobacco so frequently that it was necessary to make laws to prevent even them from using tobacco during worship.

As tobacco smoking spread through England the demand often exceeded the supply, and prices soared. London tobacco shops were equipped with balances; the buyer placed silver coins in one pan and might receive in the other pan, ounce for ounce, only as much tobacco as he gave silver. The high price, however, did not curb demand.

In 1610 an English observer noted: "Many a young nobleman's estate is altogether spent and scattered to nothing in smoke. This befalls in a shameful and beastly fashion, in that a man's estate runs out through his nose, and he wastes whole days, even years, in drinking of tobacco; men smoke even in bed." A failure of a local tobacco crop could make men desperate. Sailors approaching the island of Nias in the Malay Archipelago were greeted with cries: "Tobacco, strong tobacco. We die, sir, if we have no tobacco!"

The addictive nature of tobacco was noted at about the same time by Sir Francis Bacon, who wrote: "The use of tobacco is growing greatly and conquers men with a certain secret pleasure, so that those who have once become accustomed thereto can later hardly be restrained therefrom."

The invention of cigarette-rolling machines made the availability of tobacco even easier and was seen by some as an additional cause for alarm. The following editorial in the *Boston Medical and Surgical Journal* (1882) states the case well:

> *Our greatest danger now seems to be from an excess of cigarette smoking. The numbers of young men who smoke cigarettes is almost startling. It is not only students, but even school boys in their teens, who vigorously and openly indulge in this dangerous habit. . . . A little cigarette, filled with mild tobacco which lasts for only a few minutes, appears harmless enough. But the very ease with which these bits of paper can be lighted and smoked adds considerably to the tendency to*

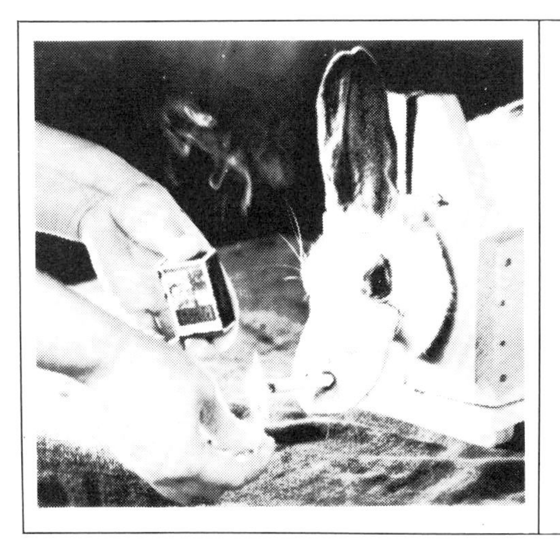

***By 1967, when this photograph was taken, this rabbit had smoked nine cigarettes a day for five years in the cause of cancer research at the Institute of Oncology in the Soviet city of Tbilisi. The rabbit came to exhibit a classic symptom of nicotine addiction, only becoming nervous when technicians were late with the cigarette.***

> *indulge to excess... One of the pernicious fashions connected with cigarette smoking is "inhaling". The ideal cigarette smoker is never so happy as when he inhales the smoke, holds it in his air passages for some time, and then blows it out in volume through nose and mouth. If he realized that "the smoker who draws the greatest amount of smoke and keeps it longest in contact with the living membrane of the air passages undoubtedly takes the largest dose of the oil", he might at least endeavor to modify his smoking in this respect.... These are dangers super-added to those attendant on the ordinary use of tobacco [other than machine rolled cigarettes], and should be considered by all medical men.*

### Current Controversy and Theory

It has long been suspected that tobacco was a drug of high abuse potential and that nicotine was the critical substance. However, the controversy has remained until recently. Some have held that nicotine was simply a toxin delivered by tobacco smoke and that it had little to do with behaviour. Others felt that nicotine actually was a noxious or aversive

***An American Cancer Society poster highlights the fact that smoking not only endangers the health of the smoker, but also reduces the quality of the smoker's immediate environment.***

element in cigarette smoke and limited how much a person smoked. Still others were convinced that nicotine itself was an abusable drug and the key to compulsive tobacco use.

Which theory was correct had implications for the understanding and treatment of cigarette smoking, as well as for government policy. For instance, should treatment be modelled after that used for people who bite their fingernails, or after that for people who abuse drugs? Should the government support the tobacco industry as it does other forms of agricultural industry, or should it regulate tobacco as a drug?

Although these are critical issues, they have not had much consideration in parliaments around the world. But in the United States the issues have been debated in both houses of Congress. Expert witnesses were found to support every conceivable position. One testified that the high incidence of cancer in cigarette smokers was due to personality traits in smokers. Representatives of the tobacco lobby argued that any regulation on cigarette smoking would be an attack on the free will of Americans to engage in voluntary pleasurable acts.

The National Institute on Drug Abuse and the United States Public Health Service testified that cigarette smoking was an instance of drug abuse. Their spokesmen insisted that cigarette smoking is not a voluntary pleasure but rather a compulsively driven behaviour.

The National Institute on Drug Abuse (NIDA) was one of the last major national or international health-related

***A Michigan resident dons a gas mask during a football game to protest against smoking in public places. The controversy over smoking in public involves two powerful lobby groups: the tobacco industry and the public health associations.***

organizations to take the official position that cigarette smoking was a form of drug abuse. The consequences of its position were potentially so widespread that NIDA had to be certain of the findings. The evidence that convinced NIDA to take this position can be summarized as follows.

### *Similarities in the Use of Tobacco and Known Drugs of Abuse*

The most obvious similarity is that, by and large, non-nutritive plant products are not widely consumed unless they contain a drug that affects the way people think or feel (psychoactivity). These substances are used in such a way that the drug gets into the blood stream and ultimately to the brain. Opium poppies are reduced and refined to yield the potent extract morphine; further processing results in heroin. Cocaine is extracted from coca leaves and processed in ways that maximize its effects.

Nicotine is made available, following an elaborate process of harvesting and manufacturing, in a very convenient and effective delivery system—the tobacco cigarette. Nicotine reaches the brain even more efficiently when inhaled in tobacco smoke than when given intravenously. It is well absorbed through the thin membranes of the mouth and

***Irene Parodi of California, U.S.A. smiles after a landmark case in which a Circuit Court of Appeals awarded her $20,000 in disability pay because she developed bronchitis after being transferred to an office where most of her co-workers smoked. Recent evidence indicates that even nonsmokers can suffer from being in the presence of smokers.***

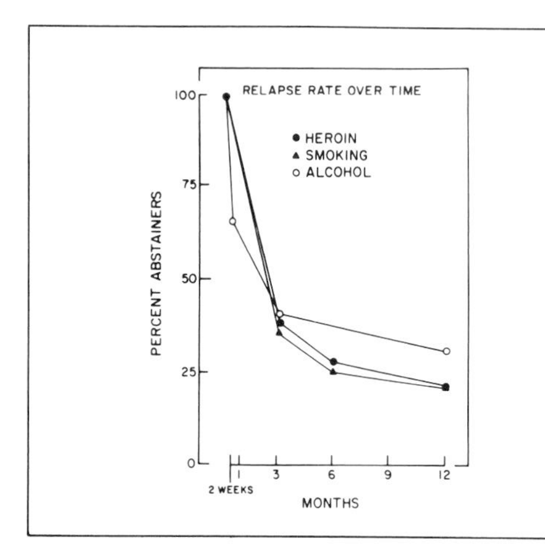


**Figure 13.** ***A survey from the early 1960s shows the rates of relapse for three drugs of abuse.***

nose and is, therefore, well absorbed when taken in the form of chewing tobacco or snuff.

Developing patterns of the behaviour of tobacco or drug use are also similar. With tobacco, alcohol, heroin, and many other drugs, initial use (experimentation) is usually with the support or even encouragement of friends or relatives. One of the best predictors of smoking, in fact, is that both parents are cigarette smokers. Similarly, a common finding among people who have recently stopped smoking is that either their friends or relatives have also stopped. These patterns of peer and social pressure are shared with most drugs of abuse.

Patterns of use also are similar. As with the opiate drugs (morphine and heroin), initial build up of tolerance leads to increased use, which then remains stable over long periods of time. Removing the drug for short periods results in discomfort and an increased desire to take the drug.

In a study and treatment programme for smokers, heroin addicts, and alcoholics, people were required to abstain from the use of the drug. Over the next year, their rate of relapse was studied. As shown in figure 13, patterns of relapse were similar for all three drugs. This suggests similar biological factors.

Other points of similarity between cigarette smoking and drug abuse are tolerance and physiological dependence. As noted earlier, tolerance to tobacco and nicotine is well documented. This effect is shared by most drugs of abuse.

The issue of physiological dependence is more complex, however. Drugs are often abused when no physiological dependence can be detected. However, many drugs are associated with discomfort and a heightened desire ("craving") to take the drug during abstinence. Cigarettes are no exception. Indeed many people will travel a fair distance to get a cigarette when they have exhausted their supply.

## *Nicotine Is an Abusable Drug*

As shown in Chapter 6 there was considerable evidence that cigarette smoking was a behaviour pattern and a form of drug abuse in which the critical drug was nicotine. However, much of this evidence was circumstantial and not considered sufficient for an agency of the United States Government to officially label cigarette smoking as a form of drug abuse. It had to be determined if nicotine itself met the criteria for being defined as a drug of abuse.

Over the years, standardized procedures were developed to discover if new drugs were likely to be abused. Two different kinds of studies were critical. One type is called "the single dose study", which compares the effects of one dose of the test drug to the effects of single doses of standard drugs of abuse. Morphine, cocaine, alcohol, and pento-

*An advertisement for* **Lucky Strike** *cigarettes proclaims the "cream of the crop" quality for their tobacco. International tobacco companies currently spend over $2 billion a year on advertising.*

barbital are all highly abused drugs that served as good standards. The other study is called the "self-administration study". This makes the drug available to both animal and human subjects to determine if they will take it voluntarily (i.e., self-administer it).

The single dose studies compared nicotine in cigarettes to nicotine given intravenously. Volunteers were given research cigarettes to smoke on some test days and were given intravenous nicotine injections on other test days. The subjects were never told what nicotine dose level they were given, or even whether the injection or cigarette contained any nicotine at all. Physiological and psychological measures were taken before and after injections or cigarettes were given.

Several critical findings were made in this study. Firstly, nicotine produced similar effects whether given intravenously or in the form of cigarette smoke. This proved that nicotine is responsible for many of the physiological and psychological effects of tobacco. Secondly, nicotine produced effects in the brain that allowed volunteers to accurately report when they had been given nicotine and when they had been given a "blank" (placebo). They were also able to rate accurately the strength of the dose. This showed that nicotine was psychoactive. Thirdly, nicotine produced psychological effects of euphoria that were similar to those produced by the standard drugs of abuse. This indicated that nicotine itself was an abusable drug that could produce compulsive behaviour. Fourthly, subjects with a wide range of experience with drugs of abuse reported that certain effects of nicotine were similar to the effects of cocaine or amphetamine.

Figure 14 illustrates one of the key findings—a psychological effect shared by nicotine and known drugs of abuse. The figure shows a comparison of nicotine to several other drugs of abuse and one drug which is not abused (zomepirac) on a "drug-liking scale". As the figure shows, volunteers can feel the drugs, they like the feelings, and the feelings are stronger with bigger doses. This psychological measure is a hallmark of abusable drugs.

To discover if the psychological effects of nicotine had to do with social and critical factors, certain studies with animals were done. Certain animals are useful for testing

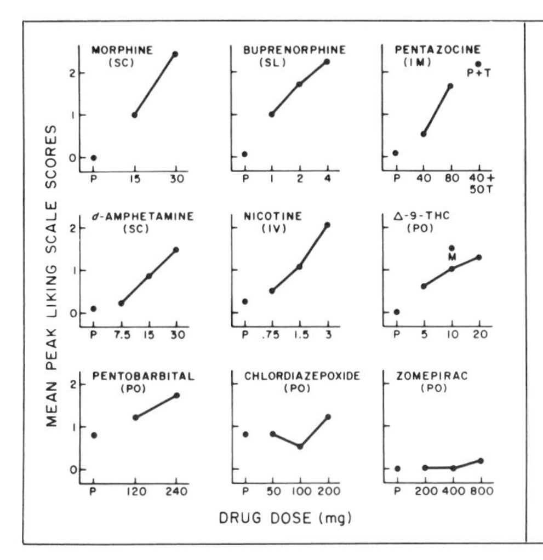


**Figure 14.** ***Average scores on a drug-liking scale are shown for volunteers tested with various drugs. SC indicates a subcutaneous injection. SL indicates a few drops placed beneath the tongue. IM indicates an intramuscular injection. IV indicates an intravenous injection. PO indicates that the drug was swallowed in capsule form. The marijuana (M on the THC graph) was smoked. Note the extraordinarily small dose of nicotine required to register a strong liking response.***

abusable drugs since they usually respond to the drugs as do humans.

With animals, however, it is generally assumed that the responses are due to the biological effects of the drugs and not to socio-cultural factors such as advertising or peer pressure. White rats were trained to press one lever when they received a stimulant and another lever when they received a sedative. When the animals were given nicotine, they pressed the stimulant lever. The larger the dose, the more they pressed the lever. Thus, like humans, the animals were reporting that they could "feel" the injections, that the injections "felt like stimulants", and they could accurately "rate" the size of the dose.

In the "self-administration studies" human volunteers and animal subjects were permitted to take nicotine intravenously using automatic injection systems. These studies also helped determine whether the drug is important by itself, or if various social and other factors are necessary for the drug to be taken. Both the human volunteers and animals voluntarily took nicotine, and nicotine was found to function as a reinforcer (reward). Figure 15 shows the pattern of nicotine self-administration by human volunteers. The patterns are interesting since they are similar to those found when people smoke cigarettes.

The self-administration studies were important in showing that nicotine could serve as a reinforcer without the other factors that accompany cigarette smoking (taste, oral

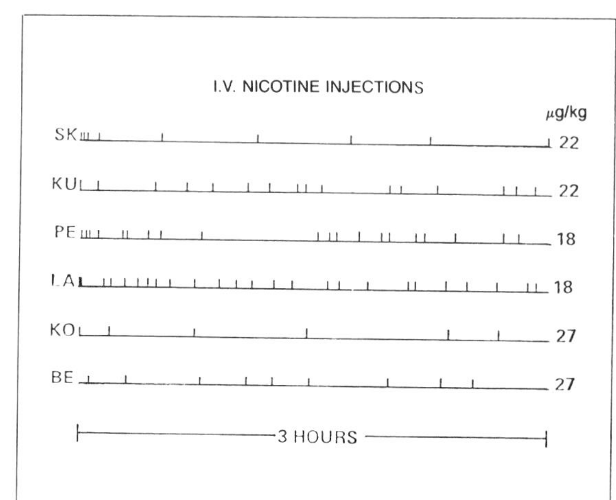


**Figure 15.** ***Patterns of nicotine injections received by six volunteers. Each dose was equal to one typical American cigarette. Actual doses in micrograms per kilogram are shown. Subjects were free to take nicotine by pressing on a lever that delivered a dose to a forearm vein. The patterns of intravenous self-administration of nicotine were very similar to those observed in people smoking cigarettes.***

satisfaction, and peer approval). The results of studies on both human and animal subjects indicate that the biological effects of nicotine are sufficient for it to serve as a reinforcer. In other words, factors unique to humans—television advertising, peer pressure, attempts to prove masculinity or independence—probably have little to do with why nicotine is a reinforcer.

Such studies do not mean that socio-cultural factors are not important. They clearly are. Rather, the studies show that nicotine's biological effects make it an ideal drug to be abused and to cause addiction, particularly when it is so widely available.

The results of these studies led the director of the National Institute on Drug Abuse (Dr. William Pollin) to testify to both the United States Senate and the House of Representatives that cigarette smoking is a form of drug abuse in which persons become dependent on the drug nicotine. In its Triennial Report to the United States Congress, the National Institute on Drug Abuse concluded the following:

> *In answer to the question, "what are the mechanisms that underlie the compulsive use of tobacco?" the National Institute on Drug Abuse is in agreement with other organizations (e.g., the American Psychiatric Association, and the World Health Organization) that tobacco use can be an*

> *addictive form of behavior. In addition to this conclusion, an appraisal of data collected by NIDA's intramural [Addiction Research Center] and extramural [e.g. Johns Hopkins and Harvard] research programs indicates that the behavior is a form of drug abuse in which nicotine is critical. Specifically, it is evident that the role of nicotine in cigarette smoking is similar to the role of cocaine in coca leaf use, of THC in marijuana smoking, and of ethanol in alcoholic beverage consumption.*

The results of these studies and of the policy statements by the National Institute on Drug Abuse and the United States Public Health Service (which adopted NIDA's policy) will take time to implement fully. However, changes of medical and political importance are already occurring. New bills have been passed in Congress that strengthen warnings on cigarette packets, and treatments for cigarette smoking are being developed based on drug abuse models.

***Major U.S. health organizations seeking legislation on smoking and public policy held a news conference in May 1984 to take up the challenge for a "smoke-free society by the year 2000."***

***A scientist studying the effects of marijuana smoke. Recent studies indicate that smoking "pot" can, like smoking cigarettes, cause lung damage.***

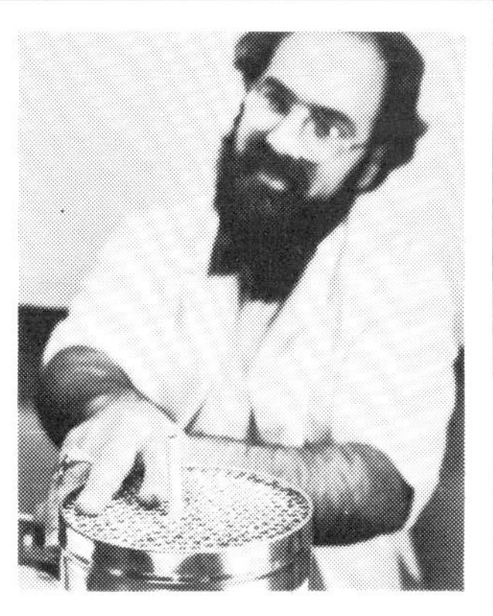

***Workers in Virginia transport the tobacco crop from the fields to the curing sheds in 1897.***

CHAPTER 8

# TOBACCO AND THE LAW

The use of tobacco has never been eliminated from a country or major culture into which it was introduced. This is noteworthy since efforts to control the use of tobacco were equal to those used to control other forms of behaviour such as crime and religious sin. Opposition to tobacco use took a variety of forms. Surprisingly, severe and immediate punishment proved no more effective in halting smoking than current fears of long-term health consequences.

Rulers came and went, but tobacco remained. Later, governments underwent a conversion of sorts, prompted primarily by the realization that tobacco was an excellent source of revenue—derived either from customs dues (such as those introduced by Cardinal Richelieu in France in 1629) or from the sale of monopolies to deal in tobacco goods. Bohemia was fortified in 1668 with money derived from the tobacco trade, and the Emperor Leopold of Austria used tobacco revenue to finance elaborate hunting expeditions. The scale of tobacco revenue is regarded by many as underlying many governments' half-hearted endorsement of the anti-tobacco cause.

## *Prohibiting Smoking*

In the early 1800s Berlin came to regard tobacco use as dangerous and contrary to the character of an orderly and civilized city. As a result, smoking was strictly forbidden, and anyone breaking the law could be arrested and punished by fine, imprisonment, or physical punishment. The prohibition was lifted briefly during a cholera outbreak in the 1830s but was restored and remained in force "because nonsmokers have a clear right not to be annoyed" until 1848. The reference to the rights of nonsmokers is echoed in the recent campaign to have greater restrictions placed on public areas in which smoking may take place.

Pope Urban VIII issued a formal decree against tobacco in 1642 and Pope Innocent X issued another in 1650, but clergy as well as laymen continued to smoke. Bavaria prohibited tobacco in 1652, Saxony in 1653, Zurich in 1667, and so on across Europe. The states, like the Church, proved powerless, however, to stem the drug's use.

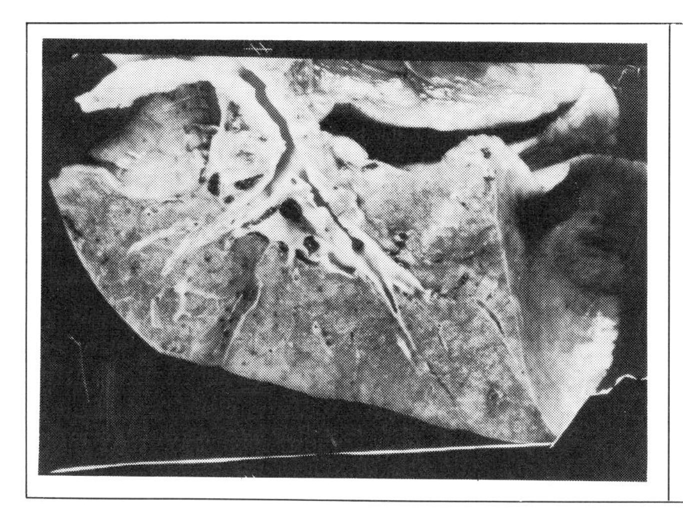

***A normal human lung, with blood vessels clearly visible, no discoloration, and no tumorous growths.***

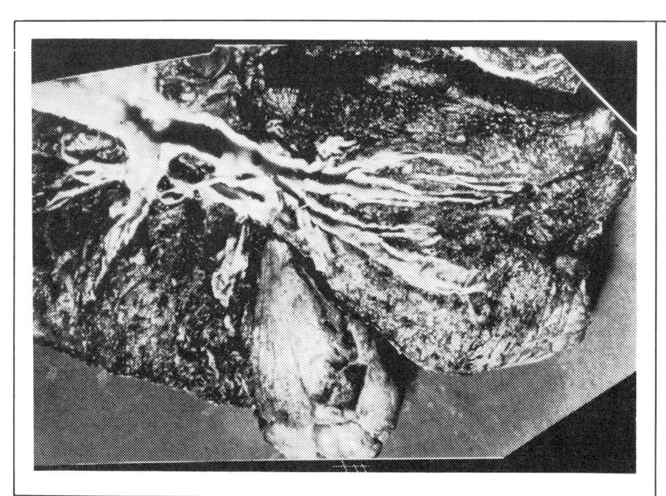

***A lung taken from a victim of emphysema, exhibiting extreme discoloration and much damaged tissue.***

In Constantinople in 1633 the Sultan Murad IV decreed the death penalty for smoking tobacco. Wherever the Sultan went on his travels or military expeditions, his stopping places were frequently marked by executions of tobacco smokers. Even on the battlefield he was fond of surprising men in the act of smoking: he would punish his own soldiers by beheading, hanging, quartering, or crushing their hands and feet and leaving them helpless between the lines. In spite of the horrors and insane cruelties inflicted by the Sultan, whose blood-lust seemed to increase with age, the passion for smoking persisted in his domain.

The first of the Romanov czars, Mikhail Feodorovich, also prohibited smoking, under dire penalties, in 1634. "Offenders are usually sentenced to slitting of the nostrils, beatings, or whippings," a visitor to Moscow noted. Yet, in 1698 smokers in Moscow would pay far more for tobacco than English smokers, "and if they lack money, they will sell their clothes for it, to the very shirt."

By 1603 the use of tobacco was well established in Japan and an edict prohibiting smoking was pronounced. As no notice was taken of the edict, more severe measures were taken in 1607 and 1609, by which the cultivation of tobacco was made a criminal offence. Finally, in 1612 it was decreed that the property of any man detected selling tobacco should be handed over to his accuser, and anyone arresting a man conveying tobacco on a pack-horse might take both

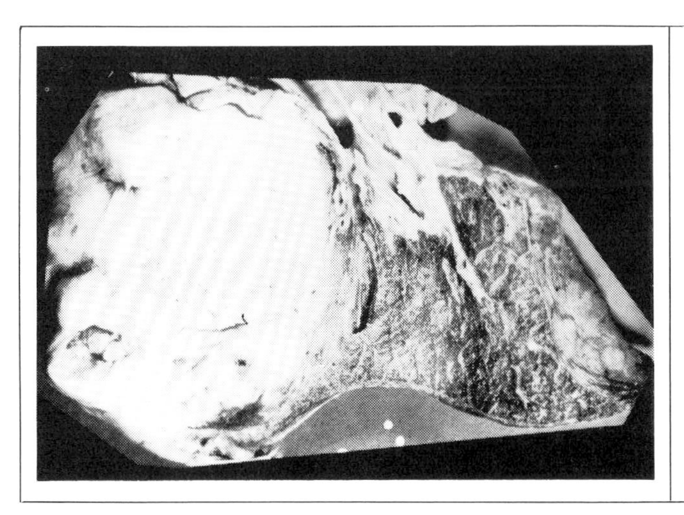

***A cancerous human lung, exhibiting extreme discoloration, shrinkage of tissue, scarring, and rampant growth of tumors.***

horse and tobacco for his own. Yet in spite of all attempts at repression smoking became so general that in 1615 even the officers in attendance on the Shogun used tobacco. The result was an even sterner measure, to the effect that anyone in the army caught smoking was liable to have his property confiscated.

It was all to no avail. The custom spread rapidly in every direction, until many smokers were to be found even in the Mikado's palace. Finally, even the princes who were responsible for the prohibition took to smoking. Tobacco had won again. In 1625 permission was given to cultivate and plant tobacco. By 1639 tobacco had taken its place as an accompaniment to the ceremonial cup of tea offered to a guest.

By 1725 even the pope was forced to capitulate. One historian has written on the subject: "Benedict XIII, who himself liked to take snuff, annulled all edicts in order to avoid the scandalous spectacle of dignitaries of the Church hastening out in order to take a few clandestine whiffs in some corner away from spying eyes."

## *Legislation and Taxation*

More recent attempts at control by legislation and taxation have had some documented effects. The average number of cigarettes sold per person, in various countries, is directly affected by price. Scandinavian countries, which have the highest tobacco taxes, have the lowest per capita use of cigarettes. Since much of the price reflects the tax on tobacco, it can be seen that heavy taxation does reduce the use of tobacco. This finding is comforting to economists and psychologists alike since it shows that in some ways tobacco is much like any other traded good. It is an "elastic" commodity, like sugar, the use of which is persistent but will vary to some degree with price.

More recently, "clean air" or "public smoking" legislation has been passed which restricts the use of tobacco in many public places. Most of these laws require certain kinds of places, such as restaurants, to maintain separate smoking and nonsmoking areas for their patrons. While such legislation has been passed to protect nonsmokers from involuntary intake of tobacco smoke, it is also seen by some as a restriction on the freedom of smokers to smoke where and when they please. The proponents of clean air ordinances argue that exposure to tobacco smoke endangers health and well-being. They claim that the right to smoke has neither a moral nor constitutional basis.

The opponents of such legislation argue that the evidence is not great enough to permit such widespread regulation of behaviour. The controversy is probably a healthy one even if smoking is not. Certainly, freedom should be protected. But the freedom of nonsmokers not to be exposed against their will to known carcinogens should be given at least equal weight with the freedom of smokers to pollute the environment. As was shown earlier, the evidence that smoking is a specific cause of death and disease is strong, much stronger than the data accepted as sufficient to eliminate lead from petrol, asbestos from building materials, and PCB's from water supplies. Although the danger to nonsmokers of breathing other people's smoke is less clear, there is an increasing amount of evidence that it can have serious adverse consequences for the health of both children and adults.

*"Don't worry. If it turns out tobacco is harmful, we can always give it up."*

***As this American Cancer Society poster makes clear, children whose parents smoke are more liable to become smokers themselves than the children of non-smokers. The natural tendency of children to use their parents as role models figures as prominently as peer pressure in the child's decision to begin smoking.***

# CHAPTER 9

# THE TREATMENT OF CIGARETTE SMOKERS

Only about one in four people who receive treatment for cigarette smoking is not smoking one year later. Most people relapse within a few months. The success rate for smoking is not much different from that for heroin addiction or alcoholism. Cigarette smoking is a form of drug abuse in which nicotine is the critical drug. As a form of drug abuse, cigarette smoking shares much in common with other forms of drug abuse.

These observations have major implications for how cigarette smokers should be treated. Hopefully, as these implications are considered they will help to improve the efficacy of treatment. The position taken in this volume, that cigarette smoking is a form of drug abuse, is based on a wealth of evidence of the overwhelming importance of the role of nicotine in smoking. It is consistent with the position taken by the World Health Organization, the International Union Against Cancer, and official bodies in the United States such as the United States Public Health Service and the National Institute on Drug Abuse.

## *Myths about Cigarette Smokers*

Our new knowledge about cigarette smoking should help to dispel some of the myths about why people smoke and why they do not stop smoking. The myths are many and include the following: cigarette smokers are ignorant of the health consequences of smoking; cigarette smokers are less intelligent than nonsmokers; cigarette smokers are psychologically self-destructive; cigarette smokers have poor moral character; cigarette smokers are weak-willed.

In point of fact, cigarette smokers are, by and large, ordinary people whose behaviour is, in part, controlled by a powerful environmental factor—nicotine. This is not an excuse for their behaviour, but rather a partial explanation for why they may *appear* to some people to be ignorant, irrational, or poorly motivated. Most people who smoke know that something about the cigarette smoke is very important to them, that they will feel genuine discomfort if they do not smoke at certain times, that they may not be able to perform well at their jobs if they are not permitted to smoke, and that there are times when they will go to great lengths to be able to smoke. Yet they are given advice by nonsmokers and therapists that include the following: "Why don't you just stop?" "It's just a matter of making up your mind." "Hide your cigarettes." "Chew gum." "Think about something else." "It's harmful so don't do it." "Take up a new hobby to distract yourself." Fortunately, the growing understanding that a powerful drug is involved should help change these views, in smokers and nonsmokers alike.

Admitting that one's behaviour is partly controlled by a drug is neither an excuse nor an admission of helplessness. It is a first, critical step in changing the behaviour of drug-taking. It is hard to treat "weak wills" and "poor moral character", but we can do something about behaviour controlled by the drug nicotine.

### *General Principles of Treatment for Smokers*

Since cigarette smoking has been seen as a form of drug abuse, new programmes of treatment have been developing rapidly. Many of these apply principles learned over the years in treating drug abuse and alcoholism. One is that the *effect of the drug in the brain is the single most critical factor* to consider. It is this effect which maintains the compulsive drug-taking behaviour of a person for years, or even a lifetime. This is a fact essential to effective treatment programmes.

As mentioned earlier, the drug of abuse can be substituted with a similar but safer drug (like nicotine chewing-gum), or the effects of the drug can be blocked (for instance, mecamylamine-like drugs can be used as a sort of immunization procedure). Alternatively, if nicotine is providing some therapeutic effects, like stress relief or weight control, then these effects might be replaced by appropriate treatment using either more appropriate drugs or non-drug treatments (for instance stress management or diet plans).

An equally important principle that has been learned from drug abuse treatment is that *non-drug factors are also important* and must be considered for effective treatment. It is probably obvious to smokers, if not to others as well, that the pleasures of smoking are many and include taste and smell, the social interactions which they make easier, and even the actual handling of the cigarettes.

Another important set of non-drug factors is the degree to which *tobacco use can become an integral part of behavioural patterns.* For instance, just as a tennis player may find it necessary to bounce the ball, or adjust his or her socks before feeling comfortable enough to serve, so too might a writer feel unready to face the typewriter without lighting up a cigarette first. The ability to function may be disrupted by the abrupt termination of smoking, and it may take years before the person "feels right" doing many things. Many writers, for instance, feel that to give up smoking would seriously impair their very livelihood. If we consider that a cigarette smoker may light 10,000 cigarettes and take 100,000 puffs per year, it should come as no surprise that smoking may become an important part of the smoker's life.

Treatment of cigarette smoking must address the possibility that *other drugs are also being abused* by the smoker. People who abuse drugs tend to abuse several. For instance, sedatives and alcohol are often abused by the same people, morphine and cocaine are often used jointly, and people who abuse any drug are likely also to smoke cigarettes (more than 90% of opiate and alcohol-dependent persons smoke cigarettes). This finding has led to treatment programmes that focus on all of the drugs abused by individuals.

The last major set of findings that have emerged from studies of drug abuse is that stopping can be accompanied by years of *sensitivity and susceptibility to relapse.* In some cases the sensitivity may be partially due to physiological changes. For instance, with the opioid drugs, a secondary and more subtle withdrawal syndrome may persist for six months or more after the initial withdrawal has subsided. Additionally, environmental stimuli may serve as cues which elicit withdrawal-associated discomfort. When this happens, relapse is more likely. In this regard, studies with both opioid drugs and tobacco have shown that stimuli associated with the respective form of drug taking (pictures of needles for the opiate users, smell of cigarette smoke for the cigarette smoker), can elicit such responses: both physiological, such as increased heart rate and sweating and psychological, such as discomfort and desire to take the drug.

Other kinds of stimuli may also enhance the likelihood of relapse. These include social situations, stress, and anxiety. In the context of drug abuse, relapse is simply one characteristic. It does not mean that treatment has failed or that the person is a failure. It simply marks a point to restore, possibly with modification, treatment.

If all of the above does not make the treatment of cigarette smoking seem like a hopeful endeavour, at least it should provide an appreciation of why success rates are generally considered good if a particular programme achieves 30% abstinence on a one-year follow-up.

### *Nicotine-Containing Chewing-Gum: Some Facts*

A nicotine-containing chewing-gum has been developed for use in the treatment of cigarette smoking. The idea was proposed by a Swedish company in the early 1970s and the gum was extensively tested in the United Kingdom and other countries before being marketed as an aid in giving up smoking. Studies showed that the proportion of smokers who could stop was doubled when they were given nicotine gum instead of a placebo. The gum is now available as a precription-only medicine in a large number of countries, including the United Kingdom, Canada, Australia, New Zealand and the United States.

Chewing the non-nutritive gum, called Nicorette®, releases nicotine, and the amount of the drug released is directly related to the rate and intensity of the chewing. The maximum amount of nicotine that may be extracted from one piece is two milligrams, approximately the amount obtained from a cigarette. The gum is buffered so that the nicotine is well absorbed through the buccal mucosae (skin in the mouth and under the tongue). Its effects can be felt within a few minutes of active chewing. While other products have been advertised as substitutes for cigarettes, this is the only one that actually is known to deliver nicotine.

The gum is a "methadone" for cigarette smokers. As such we might make a few observations: firstly, it could be the most important new advance in the treatment of cigarette smoking in decades; secondly, its effectiveness will be disappointing unless we learn from the methadone experience;

thirdly, the gum is useful because it delivers a powerful drug. Each of these observations deserves brief elaboration.

Many smokers say that they would like to stop if they could do so without excessive discomfort. The gum may be the biggest single step in giving people the freedom of that choice. Many studies have shown that the gum significantly reduces the discomfort and desire to smoke that accompany stopping. Additionally, the gum may replace some of the pleasure and benefit derived from smoking.

Just as simply giving methadone to opioid abusers without consideration of their particular problem and without supplementary treatment is not terribly effective, we can expect that similar use of the nicotine gum would produce disappointing results. The first and most obvious consideration is that the gum will probably be of little use to people who smoke only a few cigarettes. In their case the nicotine itself is relatively unimportant. Here, a simple test such as the Fagerstrom tolerance questionnaire (see Appendix 1) appears to be a useful predictor.

It is likely that persons who score high on the questionnaire (in other words, are highly dependent on nicotine itself) will be better candidates for effective treatment by the gum. There are, however, a vast array of non-nicotine factors that should be considered. For instance, it may do little good to alleviate a few days of discomfort if the person gains too much weight or becomes ineffective at work. Such people are likely to relapse. The gum should be given in conjunction with a complete programme for the treatment of smoking, perhaps tied to some of the more conventional treatments listed in the appendices.

The last point is that the gum itself should be used with caution. It contains nicotine, and as we have seen, nicotine is a powerful psychoactive drug. While it is in a form that appears safer and more manageable than the nicotine delivered in conjunction with tar, CO, cyanide, and so forth, nicotine is nonetheless a drug with its own toxic effects. Persons with cardiovascular problems, and women who are pregnant, in particular, should be aware that chewing the gum involves some of the same health risks as smoking cigarettes.

The gum should not be viewed as a long term substitute. In fact, there is the proven possibility of persons becoming dependent on the gum itself. While it can be argued that these people are probably better off than they were while smoking, clearly they are still exposing themselves to an environmental toxin.

### *Safer Cigarettes: Are They?*

Even among cigarette smokers, it has become fashionable to be "health conscious." That is, to eat right, to drink right, to exercise, and even to smoke right. As a result cigarette advertisers began to associate tobacco products with fresh air, mountain streams, sports, dance, and exercise. People, in turn, have grown more sophisticated regarding health care. They count calories, milligrams of sodium in their diets, grams of protein, and perhaps even keep track of the air quality indexes.

One way that people have grown sophisticated is that they understand that in most cases the critical factor is the *amount* of the substance which they ingest. They understand that everyone is exposed to some cancer-causing agents, for instance, but that the likelihood of getting cancer is related to the amount of those agents. Similarly, many cigarette smokers know that the hazards of smoking are related to the amount of tobacco-containing toxins that they ingest.

Unfortunately, this sophistication in the self-monitoring of what we eat, drink, and smoke has made cigarette smokers easy targets for the "safer cigarette". Although the tobacco companies do not concede that any of their products cause disease, they have nevertheless promoted low-tar brands. Average tar yields have dropped by about 50% in the past two decades, and in the United Kingdom tobacco companies have argued for the need for them to be allowed to continue cigarette advertising in order to encourage smokers to switch to low-tar brands. The official tables of tar, nicotine and carbon monoxide yields have also encouraged the concept of the less hazardous cigarette by advocating that smokers who cannot stop should switch to a lower-yielding brand.

On the surface the move towards smoking lower yielding cigarettes seems quite promising. If people are unable to give up, at least they are being exposed to smaller amounts of the toxins as cigarettes have grown steadily "safer". But *have* the cigarettes grown safer? What do the official yield figures mean?

The answer to the first question is that cigarettes are probably not safer than they were several decades ago. There is sound evidence that intake of tar and nicotine has little to do with published levels. People are choosing cigarettes based on misleading information. The problem is that official estimates have little to do with either the content of tobacco in the cigarettes or the way people actually smoke them. People can and do adjust the volume, force and frequency of puffing and inhale more or less of the smoke generated in response to the strength of the cigarette. But cigarette machines are programmed to smoke all cigarettes in exactly the same way and do not change the way they smoke in response to taste and strength factors. Clearly, testing and advertising methods must be changed if consumers are to be truthfully informed about what they are smoking.

### Precautions for Smokers Not Ready to Quit

This was the topic of a pamphlet issued by the Addiction Research Foundation of Toronto. The pamphlet made the following recommendations to lessen the medical risk of those cigarette smokers who were not yet ready to give up smoking.

1. Smoke as few cigarettes as possible no matter what their yield (studies show that people who smoke 40 to 50 cigarettes a day have been able to successfully cut back to

***Actress Brooke Shields in a U.S. Department of Health anti-smoking advertisement. Whereas during the 1930s and 1940s many celebrities appeared in advertisements claiming that smoking had contributed to their success, today's media personalities support the antismoking lobby.***

10 per day). Also avoid smoking more than two cigarettes per hour at any time of the day.

2. Smoke the lowest-yield cigarettes that you find acceptable, realizing that it may take weeks to get used to them. The greater the decrease in yields the better: differences of only 2 milligrams tar and 0.1 milligrams nicotine are too small to be important. (Note that unless no. 1 is followed, this will be a futile step.)

3. Do not block vent holes on filters.

4. Take fewer puffs per cigarette.

5. Leave longer butts (the last part of a cigarette delivers the highest yields).

6. Avoid inhaling; if you do inhale, take more shallow puffs.

7. Do not keep the cigarette in the mouth between puffs.

Additionally, the primary author of the pamphlet, Dr. Lynn Kozlowski, developed a method whereby smokers can check to see if they are inadvertently blocking ventilation holes. The method is simple. Immediately after smoking a cigarette, inspect the butt end of the cigarette. A yellow stain surrounded by a ring of clean white filter material indicates that the vent holes were open and delivering air to the mouth (and thus less smoke per puff). If the stain covers the end of the butt, then the holes were all blocked. If the stain extends to the filter on one side of the cigarette, then just the holes on that side were blocked, and so forth. Dr. Kozlowski also discovered that cigarette smokers can gradually cut back their tar and nicotine yields by monitoring the colour of the stain on the end of the filter. Darker stains indicate that more tar and nicotine were extracted and reached the mouth. Lighter stains indicate that less smoke was extracted. Smokers are encouraged to smoke their cigarettes in such a way as to leave lighter stains on the end of the filter.

### *Tips for Giving Up Smoking*

As should now be evident, good cigarette treatment programmes require much psychological and medical knowledge. However, there are some basic strategies which remain the core of modern treatment programmes. These strategies should be applied by smokers themselves to help them in their personal efforts. Since most people who stop do so without the specific aid of a treatment programme, widespread knowledge of these strategies and some of the information provided in this volume could do much to reduce cigarette smoking.

1. Discover some of the reasons why you smoke. Discussing your smoking behaviour with another person may be helpful. The questionnaire in Appendix 1 will help you to gauge your level of dependence; another questionnaire, developed by Dr. Michael Russell at the Addiction Research Unit in London, assesses the different reasons people have for their smoking (see Appendix 2).

2. Self-monitor your cigarette smoking behaviour for one or two weeks. An index card may be placed inside the wrapper of your cigarette packet. Then, each time you smoke a cigarette, you should write the time, place, and reason for smoking.

3. Steps one and two will provide you with important information about your own smoking pattern. The next

***Seeking help for a recognized addiction is the first step towards successful recovery. Self-help clinics, like this one for smokers, are among the many sources available.***

step is to cut back your cigarette intake to about 10 cigarettes per day. Make sure that all smoking that you do is according to *your* prearranged plan (for instance, only smoking in certain places and at certain times). If you do not plan, but simply "hold out as long as possible" between each cigarette, each cigarette will become of exaggerated importance and benefit.

4. After a few weeks of learning to control your smoking behaviour and smoking at reduced levels, just give it up. Studies show that gradual reduction may result in more persistent craving, even a year later. Physical discomfort from abrupt stopping following a strict ten-a-day pattern should be minimal. Try to avoid stopping at a particularly stressful time or when you may be attending a party at which cigarettes and alcohol will be available. Situations such as these may hinder your efforts.

4. For the first few days after stopping, treat yourself well. You have done one of the most important things you may ever do for yourself; you should take pride in it and reward yourself. Even if you are prone to gaining weight, this may be an appropriate time to reward your stopping smoking with good (though not a lot of ) food.

5. Seek the aid of friends and family. The biggest single correlate of both starting and stopping smoking is peers who are doing the same. Stopping with another person is a particularly useful approach. Otherwise, at least enlisting the social support of others before stopping can be very beneficial.

5. Most people who relapse do so within the first three months of stopping. During this period, in particular, try to avoid temptations (smelling a friend's tobacco smoke) and relapse situations (parties and stressful situations). Remember too that drugs used previously in conjunction with smoking may promote relapse.

6. Finally, and perhaps most importantly, if you relapse, recognize that you have not failed. Relapse is simply one normal element of breaking a pattern of behaviour in which an abusable drug is involved. The important thing is to resume your own efforts as soon as possible. Now you will have some experience and may design a better treatment for yourself in light of your new knowledge.

# APPENDIX 1

## *TOLERANCE QUESTIONNAIRE*

| ANSWER | SCORE | | |
|---|---|---|---|
| _____ | _____ | ***1.*** | How soon after you wake up do you smoke your first cigarette? |
| _____ | _____ | ***2.*** | Do you find it difficult to refrain from smoking in places where it is forbidden, e.g., in church, at the library etc? |
| _____ | _____ | ***3.*** | Which of all the cigarettes you smoke in a day is the most satisfying one? |
| _____ | _____ | ***4.*** | How many cigarettes a day do you smoke? |
| _____ | _____ | ***5.*** | Do you smoke more during the morning than during the rest of the day? |
| _____ | _____ | ***6.*** | Do you smoke if you are so ill that you are in bed most of the day? |
| _____ | _____ | ***7.*** | What brand do you smoke? |
| _____ | _____ | ***8.*** | How often do you inhale? |

### *Scoring*

The questions are scored so that higher points indicate a higher level of addiction to cigarettes. A score of 6 or more indicates a high level of physical dependence.

***1:*** One point is assigned to smoking within 30 minutes.
***2, 5, and 6:*** Items are scored with one point for yes answers.
***3:*** One point is assigned for answering "the first cigarette in the morning."
***4:*** Smokers are divided as light, 1-15 cigarettes; moderate, 16-25; and heavy, more than 25 (1 to 3 points).
***7:*** The brands are classified into three categories with low, medium, and high nicotine levels (1 to 3 points).
***8:*** Always—2 points; sometimes—1 point; never—0 points.

*From Fagerstrom, K.O., "Measuring Degree of Physical Dependence to Tobacco."* Addictive Behaviors, *vol 3., 1978, p. 235.*

# APPENDIX 2

ARU SMOKING MOTIVATION QUESTIONNAIRE

Date: . . . . . . . . .

Name: . . . . . . . . . . . Age: . . . . . . . . . . . Sex: . . . . . . . . . . .
Cigarette Consumption . . . . . . . . . . . . . . . . . . . . . . . . . .

Here are some statements about some of the reasons that people give for their smoking. Please indicate how much each statement applies to you by *drawing a circle* around the appropriate number.

| UNCERTAIN OR NOT AT ALL | A LITTLE | QUITE A BIT | VERY MUCH SO |
|---|---|---|---|
| 0 | 1 | 2 | 3 |

1. I get a definite craving to smoke when I have to stop for a while . . . . . . . . . . . . . . . . . . . . . . . . . . . . . . . 0 1 2 3
2. I light up a cigarette without realising I still have one burning in the ash tray . . . . . . . . . . . . . . . . . . . . . . . . . 0 1 2 3
3. I like a cigarette best when I am having a quiet rest 0 1 2 3
4. I get a definite pleasure whenever I smoke . . . . . . . . . . . 0 1 2 3
5. Handling a cigarette is part of the enjoyment of smoking it . . . . . . . . . . . . . . . . . . . . . . . . . . . . . . . 0 1 2 3
6. I think I look good with a cigarette . . . . . . . . . . . . . . . . 0 1 2 3
7. I smoke more when I am worried about something . . . . 0 1 2 3
8. I get a definite lift and feel more alert when smoking . . . 0 1 2 3
9. I smoke automatically without even being aware of it . . . . . . . . . . . . . . . . . . . . . . . . . . . . . . . . . . 0 1 2 3
10. I smoke to have something to do with my hands . . . . . . 0 1 2 3
11. When I run out of cigarettes I find it almost unbearable until I can get them . . . . . . . . . . . . . . . . . . . . 0 1 2 3
12. I smoke more when I am unhappy . . . . . . . . . . . . . . . . . . 0 1 2 3
13. Smoking helps to keep me going when I'm tired . . . . . . 0 1 2 3
14. I find it difficult to go as long as an hour without smoking . . . . . . . . . . . . . . . . . . . . . . . . . . . . . . . . . 0 1 2 3
15. I find myself smoking without remembering lighting up . . . . . . . . . . . . . . . . . . . . . . . . . . . . . . . . . . . . 0 1 2 3
16. I want to smoke most when I am comfortable and relaxed . . . . . . . . . . . . . . . . . . . . . . . . . . . . . . . . 0 1 2 3
17. Smoking helps me to think and concentrate . . . . . . . . . . 0 1 2 3
18. I get a real gnawing hunger to smoke when I haven't smoked for a while . . . . . . . . . . . . . . . . . . . . . . . . . . . 0 1 2 3

19. I feel I look more mature and sophisticated when smoking . . . . . . . . 0 1 2 3
20. I am very much aware of the fact when I am not smoking . . . . . . . . 0 1 2 3
21. I would find it difficult to go without smoking for as long as a week . . . . . . . . 0 1 2 3
22. I smoke to have something to put in my mouth . . . . . . . 0 1 2 3
23. I feel more attractive to the oposite sex when smoking . . . . . . . . 0 1 2 3
24. I light up a cigarette when I feel angry about something . . . . . . . . 0 1 2 3

PLEASE CHECK THAT YOU HAVE ANSWERED EVERY ITEM

## HOW TO ADD UP YOUR SCORE

Put your score for each item in the spaces next to each item number below and then add up your total score for each type of smoking.

| PSYCHO-LOGICAL IMAGE | HAND MOUTH ACTIVITY | INDULGENT | SEDATIVE | STIMULATION |
|---|---|---|---|---|
| (6) . . . . . . . . | (5) . . . . . . . . | (3) . . . . . . . . | (7) . . . . . . . . | (8) . . . . . . . . . |
| (19) . . . . . . . . | (10) . . . . . . . . | (4) . . . . . . . . | (12) . . . . . . . . | (13) . . . . . . . . . |
| (23) . . . . . . . . | (22) . . . . . . . . | (16) . . . . . . . . | (24) . . . . . . . . | (17) . . . . . . . . . |

| ADDICTIVE | AUTOMATIC | | DEPENDENCE |
|---|---|---|---|
| (11) . . . . . . . . | (2) . . . . . . . . | | (1) . . . . . . . . . |
| (18) . . . . . . . . | (9) . . . . . . . . | | (14) . . . . . . . . . |
| (20) . . . . . . . . | (15) . . . . . . . . | | (21) . . . . . . . . . |
| | | Addictive Smoking Score | . . . . . . . . . . . . |
| | | Automatic Smoking Score | . . . . . . . . . . . . |

Note: A score of 6 or more means that this type of smoking is an important motive for you. A total dependence score of 20 or more is high and suggests that you will have to be very determined if you are to succeed in giving up smoking, but it does *not* mean that you will fail.

# APPENDIX 3

*The following is an excerpt from the evidence given by Dr. Edward Brandt, Assistant Secretary of the United States Department of Health and Human Services to the U.S. Congress on March 9, 1983. It provides a useful brief summary of the health effects of smoking.*

I will begin by presenting a capsule description of the health effects of cigarette smoking and then a more detailed description of smoking and cancer and cardiopulmonary diseases. I will also address research efforts by the National Institute on Drug Abuse on the addictive characteristics of cigarette smoking.

In summary, cigarette smoking is clearly the single most important preventable cause of premature illness and death in the United States. Estimates of the number of deaths related to smoking exceed 300,000 annually. One may compare this figure with the 105,000 deaths that occur each year as a result of all injuries, 20,000 deaths from homicides, or the 40,000 infant deaths.

Cigarette smoking is one of the three major independent risk factors for coronary heart disease and arteriosclerotic peripheral vascular disease; a major cause of cancer of the lung, larynx, oral cavity, and oesophagus; and a major cause of chronic bronchitis and emphysema.

Cigarette smoking is a contributory factor in cancer of the urinary bladder, kidney, and pancreas. It is also associated with peptic ulcer disease. Maternal cigarette smoking is associated with retarded foetal growth and increased risk for spontaneous abortion, prenatal death, and slight impairment of growth and development during early childhood.

Cigarette smoking acts synergistically with oral contraceptives to enhance the probability of coronary and some cerebrovascular disease; with alcohol to increase the risk of cancer of the larnyx, oral cavity, and oesophagus; with asbestos and some other occupationally encountered substances to increase the likelihood of cancer of the lung; and with other risk factors to enhance cardiovascular risk.

Involuntary or passive inhalation of cigarette smoke can precipitate or exacerbate symptoms of existing disease states, such as asthma and cardiovascular and respiratory diseases and may be carcinogenic for nonsmokers. Smoking is also the major identifiable cause of deaths and injuries from residential fires.

Mr. Chairman, cancer was the first disease to be associated with cigarette smoking. As Dr. Koop pointed out in introducing our 1982 report on smoking and health a few weeks ago, reports linking smoking and lung cancer began appearing in the scientific literature as long as 50 years ago. In 1964, when the Surgeon General's Advisory Committee's report was issued, lung cancer in men, and chronic bronchitis in both men and women, were the two diseases which the committee identified as being caused by cigarette smoking.

The evidence which links cigarette smoking with lung and other cancers was reviewed in the most careful detail in the 1982 report just issued. Today, 18 years after the 1964 report, additional human experience and enormous amounts of new research make it possible for science to conclude that cigarette smoking is a major cause of cancers of the lung, larynx, oral cavity, and oesophagus, and that it is a contributory factor in the development of cancers of the bladder, pancreas, and kidney. Lung cancer accounts for one out of every four cancer deaths, and 85% of these are due to smoking. Overall approximately 30% of *all* cancer deaths are attributable to tobacco use.

A subject which was hardly touched upon in the 1964 report is the effect of smoking on women, and in the case of maternal smoking, its effect on the foetus and infant. In 1980 this was the topic of the department's report to Congress. Its conclusions were that women are not immune to the damaging effects of smoking and that the lesser occurrence of smoking-related diseases among women smokers is a result of women having lagged one-quarter century behind men in their widespread use of cigarettes.

The 1980 report established that cigarette smoking is a major threat to the outcome of pregnancy and the well-being of the baby. The risk of spontaneous abortion, foetal death, and neonatal death increases directly with increasing levels of maternal smoking during pregnancy. Smoking causes a markedly increased risk of heart attack and haemorrhage.

Another public health question, now enormously important, relates to the use of the new, low-yield cigarettes. This was the subject of the department's 1981 report. The report's conclusions were that although there is no safe cigarette, smoking cigarettes with lower yields of tar and nicotine poses a lower risk of lung cancer than smoking higher-yield cigarettes, provided there is no compensatory change in smoking patterns. Increasingly, smokers have turned to these lower-yield products; there is evidence to suggest that in doing so, at least some have increased their smoking or changed the way they smoke. This may have negated any potential benefit in their having switched to these products.

***Tobacco has been an important trade commodity since traders first introduced it to Europe in the 16th century.***

# APPENDIX 4

## *RESEARCH STRATEGIES IN CIGARETTE SMOKING*

Since this volume is written from the perspective of science, it is important to explain the general philosophy—especially the nature of scientific proof. In other words, what do we mean when we say we have proved something or that one thing causes another.

Science is basically a system for discovering lawful relations in the universe. It may consist of searching for a deadly flu virus or discovering the path of a far-off asteroid. The methods are basically the same. Initially we observe and carefully record the phenomenon and all that seems relevant. This was done for cigarette smoking in a study called the Multiple Risk Factor Intervention Trial (MRFIT). The MRFIT study is the most recent and reliable of such studies. It recorded a wide range of health-related behaviours, such as cigarette smoking and diet, in nearly 13,000 people selected from a pool of nearly 400,000 volunteers. They were divided into two groups. One group was the Special Intervention Group. These people were given counselling for ten weeks on various health-related behaviours including diet, exercise, and stopping cigarette smoking (''test'' or ''experimental'' group). The other group was comprised of similarly screened individuals (same average age, education level, etc.), but this group received no treatment ("control" group). All volunteers were given extensive medical examinations over the next six years. Over this six-year period, the occurrences of death and disease in these two groups were recorded. This study showed that health-related behaviours such as stopping cigarette smoking had a substantial effect in reducing death and disease. These findings do not prove cause and effect, but they strongly suggest it, and they give us some hypotheses to test. In other words, such data give us some rational basis for further investigations. They also tell us the probability of certain events. Probabilities can be very important in and of themselves, as any gambler knows. The probability of winning a certain bet will influence where and how much money is gambled. Why the odds are at certain levels need be of little interest. Similarly, if we discover that cigarette smoking during pregnancy is associated with a two-fold increase in the risk of birth defects, then at least some women will stop smoking during pregnancy. Much of what is called epidemiologic data is of this nature. The main point to remember is that properly conducted studies can yield important findings, even though the relations are just probabilities.

These kinds of studies can take us one step beyond simply telling the odds. If enough of these studies are conducted, a cause-and-effect relation may become so obvious that people will say it has been proven. The key is that when many bits of epidemiologic data fit a puzzle, the total picture can be conclusive.

The other scientific approach is the experimental method, a powerful scientific strategy in which all possible factors of relevance are controlled. Then the factor of interest is tested (the "independent variable"). At the same time all possible effects of interest (the "dependent variables") are measured. With the experimental method, cause-and-effect relations are clearer than in historical, observational, and epidemiologic approaches. When it has been shown that manipulation of a particular independent variable is reliably followed by a particular change in the dependent variable, we may assume that a causal relationship exists between them. For instance, to discover what parts of cigarette smoke cause cancer, the smoke is reduced and various elements such as tar and nicotine are extracted. Then solutions containing tar or nicotine can be placed on the skin of animals to see if cancer growth results. If it does, future experiments may study factors thought to be important. For instance, *dose.* By applying the solution in different strengths, it can be learned if the cancer is worse or more likely. In fact, a "safe" level, below which cancer does not occur, might even be discovered.

# APPENDIX 5

Some useful addresses for those who want further information or literature.

*United Kingdom*

ASH (Action on Smoking and Health) was set up as an anti-smoking pressure group. Although run on a shoestring budget it is highly effective.

ASH
5 – 11 Mortimer Street
London. W1N 7RH.

*Ash Branches*
*SCOTTISH ASH*

Mrs Alison Hillhouse
Director
c/o Royal College of Physicians
9 Queen Street
EDINBURGH EH2 1JQ

*ASH IN WALES*

Ms Naomi King
Secretary
92 Cathedral Road
CARDIFF CF1 9PH

*NORTHERN IRELAND ASH*

Mr Michael Wood
Director
Ulster Cancer Foundation
40 Eglantine Avenue
BELFAST BT9 6DX

*NORTHERN ASH*

Ms Irene Coghill
Secretary
Brunton Park Health Centre
Princes Road
NEWCASTLE UPON TYNE NE3 5NF

*NORTHWEST ASH*

Mrs Patricia Jones
Secretary
The Uplands
Bury New Road
Whitefield
MANCHESTER

*NORFOLK ASH*

Ms Joanna Vincent
12 The Avenue
NORWICH NR2 3PH

*SUFFOLK ASH*

Mr Ron Rossington
17 The Close
Tattingstone
IPSWICH

*SOLENT ASH*

Mrs Christine Hay
Secretary
Community Health Department
Dunsbury Way
Leigh Park
HAVANT HANTS

*WEST MIDLANDS ASH*

Mr J W Urwin
Secretary
Community Health Services
George House
St Johns Square
WOLVERHAMPTON

*YORKSHIRE ASH*

Mrs Catherine Woodhouse
Hollin Corner
65 Parish Ghyll Lane
ILKLEY LS29 9QP

*GRAMPIAN ASH*

Mrs Maisie Abbott
Honorary Secretary
"Birchwood"
Bartle Moor
Kincardine O'Neil
AB3 5EE

*CLEVELAND ASH*

Mrs Carole Johnson
Health Education Centre
West Lane Hospital
MIDDLESBOROUGH TS5 4EE

The Health Education Council acts as a focus for anti-smoking publicity campaigns and distributes literature on the effects of smoking and how to give up.

Health Education Council
78 New Oxford Street
London. WC1A 1AH

## *Further Reading*


Ashton, H. and Stepney, R. *Smoking, Psychology, and Pharmacology.* London: Tavistock Publications, 1982. Provides a good overview of the history of cigarette smoking, interesting historical anecdotes, and a summary of many of the critical factors involved in the behaviour of smoking.

Austin, G.A. *Perspectives on the History of Psychoactive Substance Use.* NIDA Research Issues, no. 24. Washington, D.C.: U.S. Government Printing Office, 1978. A detailed chronological listing of historical events concerning the discovery and spread of tobacco throughout the world.

Brecher, E.M. and Consumer Reports. *Licit and Illicit Drugs.* Boston: Little, Brown, 1972. Includes a chapter on cigarette smoking that contains interesting anecdotes and relates cigarette smoking to other forms of drug abuse.

Gilman, A.G., Goodman, L.S. and Gilman, A. *The Pharmacological Basis of Therapeutics.* New York: Macmillan, 1980. A professional level text that provides detailed information on the effects of nicotine on the brain.

Grabowski, J. and Bell, C.S. (eds.). *Measurement in the Analysis and Treatment of Cigarette Smoking Behavior.* NIDA Research Monograph, no. 48. Washington, D.C.: U.S. Government Printing Office, 1983. Provides information relating to all aspects of measuring cigarette smoking behaviour and biological intake of tobacco products.

Henningfield, J.E. "Behavioral Pharmacology of Cigarette Smoking". In T. Thompson and P. B. Dews (eds). *Advances in Behavioral Pharmacology*, vol. 4, 1984. This chapter provides a detailed scientific analysis of cigarette smoking behaviour.

Jacobson, B. *The Ladykillers: Why Smoking is a Feminist Issue.* London: Pluto Press 1981. A provocative account written for the non-specialist of the problems smoking poses for women.

Krasnegor, N.A. (ed.). *Cigarette Smoking as Dependence Process.* NIDA Research Monograph, no. 23. Washington, D.C.: U.S. Government Printing Office, 1979. Contains several chapters which address cigarette smoking as a possible form of drug abuse.

Krasnegor, N.A. (ed.). *The Behavioral Aspects of Smoking.* NIDA Research Monograph, no. 26. Washington, D.C.: U.S. Government Printing Office, 1979. Contains several chapters which address the behavioural analysis and treatment of cigarette smoking.

Kozlowski, L.T. *Tar and Nicotine Ratings May Be Hazardous to Your Health: Information for Smokers Who Are Not Ready to Stop.* Ontario: Addiction Research Foundation, 1982. A consumer-level pamphlet which describes how FTC ratings of cigarette tar and nicotine are obtained and what they mean.

National Institute on Drug Abuse. *First Triennial Report to the U.S. Congress on Drug Dependence Research.* NIDA Research Monograph. Washington, D.C.: U.S. Government Printing Office, 1984, in press. States the official position of the National Institute on Drug Abuse, that cigarette smoking is a form of drug abuse, and a summary of the recent data supporting that position.

Pomerleau, O.F. and Pomerleau, C.S. *Break the Smoking Habit: A Behavioral Program for Giving Up Cigarettes.* Champaign, Ill.: Research Press, 1977. An excellent self-help book with a systematic programme for stopping and many tips for maintaining abstinence.

Royal College of Physicians. *Health or Smoking?* London: Pitman 1983. The most recent in an authoritative series of reports on the health effects of smoking.

Russell, M.A.H. "Tobacco Smoking and Nicotine Dependence". In: Gibbins, R.J., Isreal, Y., Kalant, H., Popham, R.E., Schmidt, W., and Smart, R.G. (eds.). *Research Advances in Alcohol and Drug Problems*, vol. 3. New York: Wiley, 1976. An early but important technical review of the pharmacology and psychology of cigarette smoking.

Surgeon General. *Smoking and Health.* Washington, D.C.: U.S. Government Printing Office, 1979. *The Health Consequences of Smoking: The Changing Cigarette.* Washington, D.C.: U.S. Government Printing Office, 1981. Provides a comprehensive review of the toxins contained in cigarettes, the medical effects of smoking and trends in smoking in the United States. Other Surgeon General's Reports address topics such as smoking and women, smoking and cancer, and smoking and heart disease.

Taylor, P. *Smoke Ring: The Politics of Tobacco.* London: Bodley Head 1984. A fascinating account of how governments' will to act effectively against smoking is paralysed by their involvement with a huge multinational industry.

*Tobacco Reporter.* Raleigh, N.C. The trade journal of the tobacco industry provides an interesting perspective of the marketing strategies, economic issues, and political issues concerning tobacco manufacturing.

*Why People Smoke Cigarettes.* Publication No. 83-501195, Washington, D.C.: Public Health Service, 1983. A consumer-level pamphlet stating the position of the United States Public Health Service, that cigarette smoking is a form of drug abuse.

# *Index*

***Dr. Jack E. Henningfield*** received a Ph.D. in psychopharmacology from the University of Minnesota. He is an assistant professor of behavioral biology at The Johns Hopkins University School of Medicine. He is also a consultant to The Office on Smoking and Health at the Addiction Research Center of the National Institute on Drug Abuse.

***Malcolm Lader,*** D.Sc., Ph.D., M.D., F.R.C. Psych. is Professor of Clinical Psychopharmacology at the Institute of Psychiatry, University of London, and Honorary Consultant to the Bethlem Royal and Maudsley Hospitals. He is a member of the External Scientific Staff of the Medical Research Council. He has researched extensively into the actions of drugs used to treat psychiatric illnesses and symptoms, in particular the tranquillizers. He has written several books and over 300 scientific articles. Professor Lader is a member of several governmental advisory committees concerned with drugs.

***Martin Jarvis,*** M.A., B.Sc., M.Phil. is a Lecturer in Psychology at the Institute of Psychiatry, University of London and Honorary Senior Clinical Psychologist with the Bethlem Royal and Maudsley Hospitals. He is based at the Addiction Research Unit and has researched and published extensively in the field of the addictions.